大数据与人工智能技术丛书

大数据专业英语教程

◎ 张强华 刘俊辉 郑聪玲 司爱侠 编著

清华大学出版社

北京

内 容 简 介

本书是大数据专业英语教材，内容包括大数据基础、软件与开发技术、操作系统、Python 与 R 编程语言、数据结构、数据库与数据仓库、云存储与数据备份、数据处理与数据清洗、数据挖掘、Hadoop 与 Spark、数据可视化、大数据安全等。

本书体例新颖，适合教学。每个单元均包含以下部分：课文——选材广泛、风格多样、切合实际的两篇专业文章；单词——给出课文中出现的新词，读者由此可以积累大数据专业的基本词汇；词组——给出课文中的常用词组；缩略语——给出课文中出现的、业内人士必须掌握的缩略语；难句讲解——讲解课文中出现的疑难句子，分析其语法结构，培养读者的阅读理解疑难句子的能力；习题——既有针对课文的练习，也有一些开放性的练习；短文翻译——培养读者的翻译能力；参考译文——让读者对照理解以提高翻译能力。

本书吸纳了作者近 20 年的 IT 行业英语翻译与图书编写经验，与课堂教学的各个环节紧密结合，支持备课、教学、复习及考试各个教学环节，有配套的 PPT、参考答案等。

本书既可作为高等本科院校、高等专科院校大数据相关专业的专业英语教材，也可供从业人员自学；作为培训班教材，亦颇得当。

本书封面贴有清华大学出版社防伪标签，无标签者不得销售。
版权所有，侵权必究。举报：010-62782989，beiqinquan@tup.tsinghua.edu.cn。

图书在版编目(CIP)数据

大数据专业英语教程/张强华等编著.—北京：清华大学出版社，2019（2024.8重印）
（大数据与人工智能技术丛书）
ISBN 978-7-302-52692-6

Ⅰ. ①大… Ⅱ. ①张… Ⅲ. ①数据处理–英语–教材 Ⅳ. ①TP274

中国版本图书馆 CIP 数据核字（2019）第 057436 号

策划编辑：魏江江
责任编辑：王冰飞　李　晔
封面设计：刘　键
责任校对：李建庄
责任印制：沈　露

出版发行：清华大学出版社
网　　址：https://www.tup.com.cn，https://www.wqxuetang.com
地　　址：北京清华大学学研大厦A座　　邮　编：100084
社 总 机：010-83470000　　邮　购：010-62786544
投稿与读者服务：010-62776969，c-service@tup.tsinghua.edu.cn
质量反馈：010-62772015，zhiliang@tup.tsinghua.edu.cn
课件下载：https://www.tup.com.cn，010-83470236

印 装 者：大厂回族自治县彩虹印刷有限公司
经　　销：全国新华书店
开　　本：185mm×260mm　　印　张：16　　字　数：367 千字
版　　次：2019 年 7 月第 1 版　　印　次：2024 年 8 月第 6 次印刷
印　　数：6501～7300
定　　价：39.80 元

产品编号：079685-01

前　言

　　党的二十大报告指出：教育、科技、人才是全面建设社会主义现代化国家的基础性、战略性支撑。必须坚持科技是第一生产力、人才是第一资源、创新是第一动力，深入实施科教兴国战略、人才强国战略、创新驱动发展战略，开辟发展新领域新赛道，不断塑造发展新动能新优势。高等教育与经济社会发展紧密相连，对促进就业创业、助力经济社会发展、增进人民福祉具有重要意义。

　　我们正在从信息技术时代进入数据技术时代。我国的大数据产业已经进入高速发展期，许多高校都开设了大数据专业，培养急需的专业人员。由于大数据产业有极高的发展速度，从业人员必须掌握许多新技术、新方法，因此对专业英语要求较高。具备相关技能并精通专业外语的人员往往会赢得竞争，成为职场中不可或缺的核心人才与领军人物。

　　本书的特点与优势如下：

　　（1）选材全面，包括大数据基础、软件与开发技术、操作系统、Python与R编程语言、数据结构、数据库与数据仓库、云存储与数据备份、数据处理与数据清洗、数据挖掘、Hadoop与Spark、数据可视化、大数据安全等。书中许多内容非常实用，具有广泛的覆盖面。

　　（2）体例新颖，非常适合教学，与课堂教学的各个环节紧密结合，支持备课、教学、复习及考试各个教学环节。每个单元均包含以下部分：课文——选材广泛、风格多样、切合实际的两篇专业文章；单词——给出课文中出现的新词，读者由此可以积累大数据专业的基本词汇；词组——给出课文中的常用词组；缩略语——给出课文中出现的、业内人士必须掌握的缩略语；难句讲解——讲解课文中出现的疑难句子，分析其语法结构，培养读者的阅读理解疑难句子的能力；习题——既有针对课文的练习，也有一些开放性的练习；短文翻译——培养读者的翻译能力；参考译文——让读者对照理解以提高翻译能力。

　　（3）习题量适当，题型丰富，难易搭配，便于教师组织教学。

　　（4）教学支持完善，有配套的PPT、参考答案等。

　　（5）作者有近20年IT行业英语图书的编写经验。在作者编写的英语书籍中，有三部国家级"十一五"规划教材，一部全国畅销书，一部获华东地区教材二等奖图书。基于这些图书的编写经验有助于本书内容的完善与提升。

　　在使用本书的过程中，有任何问题都可以通过电子邮件与我们交流，我们一定会给予答复。邮件标题请注明姓名及"索取大数据英语参考资料"字样。我们的E-mail地址为zqh3882355@sina.com和zqh3882355@163.com。

如本书有任何不妥之处，望大家不吝赐教，让我们共同努力，使本书成为一部"符合学生实际、切合行业实况、知识实用丰富、严谨开放创新"的优秀教材。

作　者

目 录

在线音频

Unit 1 1

  Text A Big Data 1

      New Words 4

      Phrases 6

      Abbreviations 7

      Notes 7

      Exercises 8

  Text B Big Data Analytics 10

      New Words 13

      Phrases 14

      Abbreviations 15

      Exercises 15

  参考译文   大数据 15

Unit 2 18

  Text A Computer Software 18

      New Words 22

      Phrases 23

      Notes 24

      Exercises 25

  Text B Software Development Process 27

      New Words 30

      Phrases 31

      Abbreviations 32

      Exercises 32

  参考译文   计算机软件 33

Unit 3 37

  Text A Operating System 37

New Words		41
Phrases		42
Abbreviations		43
Notes		43
Exercises		44
Text B	ETL	46
New Words		51
Phrases		53
Abbreviations		54
Exercises		55
参考译文	操作系统	55

Unit 4 58

Text A	R Programming Language	58
New Words		61
Phrases		64
Abbreviations		65
Notes		65
Exercises		66
Text B	Python Programming Language	69
New Words		74
Phrases		76
Abbreviations		77
Exercises		77
参考译文	R 编程语言	78

Unit 5 81

Text A	Data Structure	81
New Words		83
Phrases		84
Abbreviations		84
Notes		85
Exercises		86
Text B	Structured Data, Semi-structured Data and Unstructured Data	88
New Words		93
Phrases		94
Abbreviations		95

　　　　Exercises ··· 96
　　参考译文 数据结构 ··· 96

Unit 6 ·· 99

　　Text A　Basic Concepts of Database ·· 99
　　　　New Words ··· 102
　　　　Phrases ·· 104
　　　　Abbreviations ·· 104
　　　　Notes ·· 104
　　　　Exercises ··· 106
　　Text B　How Cloud Storage Works ··· 108
　　　　New Words ··· 111
　　　　Phrases ·· 111
　　　　Exercises ··· 112
　　参考译文 数据库基本概念 ··· 113

Unit 7 ·· 116

　　Text A　Data Warehouse Frequently Asked Questions ·· 116
　　　　New Words ··· 119
　　　　Phrases ·· 120
　　　　Abbreviations ·· 121
　　　　Notes ·· 121
　　　　Exercises ··· 122
　　Text B　Data Backup ··· 124
　　　　New Words ··· 128
　　　　Phrases ·· 129
　　　　Abbreviations ·· 129
　　　　Exercises ··· 129
　　参考译文 数据仓库常见问题 ··· 130

Unit 8 ·· 133

　　Text A　Data Preprocessing ··· 133
　　　　New Words ··· 136
　　　　Phrases ·· 138
　　　　Abbreviations ·· 139
　　　　Notes ·· 139
　　　　Exercises ··· 140

Text B　Data Cleansing ··· 142
　　　　　New Words ··· 149
　　　　　Phrases ·· 151
　　　　　Exercises ·· 152
　　参考译文　数据预处理 ·· 152

Unit 9 ··· 156
　　　Text A　Data Mining ··· 156
　　　　　New Words ··· 160
　　　　　Phrases ·· 162
　　　　　Notes ··· 163
　　　　　Exercises ·· 164
　　　Text B　Top 6 Data Mining Algorithms ······································ 166
　　　　　New Words ··· 175
　　　　　Phrases ·· 177
　　　　　Abbreviations ·· 177
　　　　　Exercises ·· 178
　　参考译文　数据挖掘 ··· 178

Unit 10 ··· 182
　　　Text A　What is Hadoop? ·· 182
　　　　　New Words ··· 185
　　　　　Phrases ·· 186
　　　　　Abbreviations ·· 186
　　　　　Notes ··· 187
　　　　　Exercises ·· 188
　　　Text B　Apache Spark ·· 191
　　　　　New Words ··· 195
　　　　　Phrases ·· 195
　　　　　Abbreviations ·· 196
　　　　　Exercises ·· 197
　　参考译文　什么是Hadoop ··· 197

Unit 11 ··· 200
　　　Text A　Data Visualization ··· 200
　　　　　New Words ··· 207
　　　　　Phrases ·· 209

　　　　　Abbreviations ·· 210
　　　　　Notes ··· 211
　　　　　Exercises ··· 212
　　Text B　The 14 Best Data Visualization Tools ·· 216
　　　　　New Words ··· 220
　　　　　Phrases ·· 221
　　　　　Abbreviations ·· 222
　　　　　Exercises ··· 222
　　参考译文　数据可视化 ··· 223

Unit 12 ··· 228

　　Text A　How to Manage Big Data's Big Security Challenges ····················· 228
　　　　　New Words ··· 231
　　　　　Phrases ·· 232
　　　　　Abbreviations ·· 232
　　　　　Notes ··· 233
　　　　　Exercises ··· 233
　　Text B　The Future of Big Data—Big Data 2.0 ··· 236
　　　　　New Words ··· 239
　　　　　Phrases ·· 240
　　　　　Abbreviations ·· 240
　　　　　Exercises ··· 240
　　参考译文　如何管理大数据的大安全挑战 ··· 241

Unit 1

Text A

Big Data

Big data is changing the way people work together within organizations. It is creating a culture in which business and IT leaders must join forces to realize the value from all data. Insights from big data can enable all employees to make better decisions — deepening customer engagement, optimizing operations, preventing threats and fraud, and capitalizing on new sources of revenue.

1. The Big Vs

1.1 Value

This is indeed the holy grail of big data and what we are all looking for. One has to demonstrate value that can be extracted from big or small data in order to justify the investments, whether on big data or on traditional analytics, data warehouse or business intelligence tools, whatever may be the buzzing nomenclature. There seems to be an increasing interest related to the value of big data, as indicated by the number of Google searches looking for similar terms over the last two years.

1.2 Volume

There is no doubt that the information explosion has redefined the connotation of volumes. There are several such staggering statistics going around and it has become increasingly difficult to keep track of the number and magnitude of the prefixes attached to "bytes" while measuring the volume. Since there is a "helluva lot of data", the term "Hellabyte" has been coined beyond Petabytes, Exabytes, Zettabytes and Yottabytes. However, since these measures will be superseded by the likes of Brontobytes, Geopbytes and more, lets move on!

1.3 Velocity

Similarly, velocity refers to the speed at which the data is generated. Some of the factors that exacerbate this trend are the proliferation of social media and the explosion of IoT (Internet of Things). In the context of business operations that have not yet been touched by social media or IoT, the velocity arises from sophisticated enterprise applications that capture each and every minute detail involved in the completion of a particular business process. Enterprise applications have traditionally captured such information but the world has woken up to the power of such information largely in the big data era.

1.4 Variety

The last of the original attributes of big data is variety. Since we are living in an increasingly digital world where technology has invaded into our glasses and watches, the variety of data that is generated is mind-boggling. The computing power available is able to process unstructured text, images, audio, video and data from sensors in the IoT (Internet of Things) world that capture (almost) everything around us. This attribute of big data is more relevant today than it ever was.

1.5 Veracity or Validity

Veracity or validity of data is extremely important and fundamental to the extraction of value from the underlying data. Veracity implies that the data is verifiable and truthful. If this condition is violated, the results can be catastrophic. More importantly, there are several cases in which the data is accurate but may not be valid in the particular context. For instance, if we are trying to ascertain the volume of searches on Google related to big data, we will also obtain results pertaining to the hit single "dangerous" from "big data".

1.6 Visible

Information silos have always existed within enterprises and have been one of the major roadblocks in the attempt to extract value from data. Relevant information should not only exist, but also be visible to the right person at the right time. Actionable data needs to be visible transcending the boundaries of functions, departments and even organizations for value unlocking. Individuals might have believed that information in their hands is power but in the age of big data, collective information available to the world at large is truly omnipotent!

1.7 Visual

We live in an increasingly visual world and the statistics of increase in the number of images and videos shared on the Internet is staggering. According to official statistics, 300 hours of video are uploaded every minute on YouTube. In a business context, appropriate visualization of data is critical for the management to be able to extract value from their limited time, resources and even more limited attention span!

2. More Contenders

In addition to the 7 V's described above, there are several other V's that may be considered:

2.1 Volatility

With more applications such as SnapChat and IoT sensors, we may have data in and out in a snap. Volatility of the underlying data sources may become one of the defining attributes in the future.

2.2 Variability

One of the cornerstones of traditional statistics is standard deviation and variability. Whether or not it makes to an extended list of V's relating to big data, it can never be ignored.

2.3 Viability

Embedded in the concept of value is the need to check the viability of any project. Big data projects can scale up to gigantic proportions and guzzle a lot of resources very quickly. Those who do not learn this fast and get fascinated with fads will funnel funds towards futility resulting in failure. In a nutshell, viability of any project needs to be established and big data projects do not have the liberty of exemption, whether or not it remains a trending buzzword.

2.4 Vitality

Vitality or criticality of the data is another concept that is crucial and is embedded in the concept of Value. Information that is more meaningful or critical to the underlying business objective needs to be prioritized. Analysis paralysis needs to be replaced with a more pragmatic approach. Technology allows marketers to create segments of one, but is such extreme segmentation vital or even aligned to the organizational strategy?

2.5 Vincularity

Derived from Latin, it implies connectivity or linkage. This concept is very relevant in today's connected world. There is significant value arbitrage potential by connecting diverse information sets. For instance, the government has forever been trying to connect the details of major expenditure heads and correlating the same with the income declared in tax returns to identify concealment of income. The same purpose may now be achieved by drawing information from social media posts.

3. An Example of Big Data

An example of big data might be petabytes (1,024 terabytes) or exabytes (1,024 petabytes) of data consisting of billions to trillions of records of millions of people — all from different sources (e.g. Web, sales, customer contact center, social media, mobile data and so on). The data is typically loosely structured data that is often incomplete and inaccessible.

New Words

realize	[ˈriəlaiz]	vt.认识到，了解，实现，实行
engagement	[inˈgeidʒmənt]	n.参与度，敬业度
fraud	[frɔːd]	n.欺骗，欺诈行为
indeed	[inˈdiːd]	adv.真正地，确实；当然
demonstrate	[ˈdemənstreit]	vt.示范，证明，论证
nomenclature	[nəˈmənkletʃə]	n.系统命名法；命名；术语；专门名称
analytics	[ˌænəˈlitiks]	n.分析学，解析学，分析论
redefine	[ˌriːdiˈfain]	v.重新定义
connotation	[ˌkɔnəuˈteiʃən]	n.内涵
staggering	[ˈstægəriŋ]	adj.令人惊愕的，难以置信的
helluva	[ˈheləvə]	adj.很大的
Hellabyte	[ˈheləbait]	n.数据单位，$=10^{27}$ Byte
Exabyte	[ˈeksəbait]	n.数据单位，缩写为 EB
Zettabyte	[ˈzetəbait]	n.数据单位，缩写为 ZB
Yottabyte	[ˈjɔtəbait]	n.数据单位，缩写为 YB
Brontobyte	[ˈbrɔntəbait]	n.数据单位，缩写为 BB
Geopbyte	[ˈdʒiəpbait]	n.数据单位，缩写为 GB
velocity	[viˈlɔsiti]	n.高速性；速度，速率
exacerbate	[eksˈæsəbeit]	vt.使恶化，使加剧
trend	[trend]	n.倾向，趋势

单词	音标	释义
proliferation	[prəuˌlifəˈreiʃən]	n.增殖；扩散
explosion	[iksˈpləuʒən]	n.爆发，爆炸
era	[ˈiərə]	n.时代，纪元，时期
variety	[vəˈraiəti]	n.多样性；品种，种类
attribute	[əˈtribju(:)t]	n.属性，品质，特征
mind-boggling	[maind-ˈbɔgliŋ]	adj.令人难以置信的
unstructured	[ʌnˈstrʌktʃəd]	adj.非结构化的，未组织的
veracity	[vəˈræsiti]	n.真实性
validity	[vəˈliditi]	n.有效性；合法性，正确性
extremely	[iksˈtri:mli]	adv.极端地，非常地
fundamental	[ˌfʌndəˈmentəl]	adj.基础的，基本的 n.基本原则，基本原理
verifiable	[ˈverifaiəbl]	adj.能证实的
truthful	[ˈtru:θful]	adj.诚实的，说实话的
violate	[ˈvaiəleit]	vt.违犯，冒犯，干扰；违反
catastrophic	[ˌkætəˈstrɔfik]	adj.悲惨的，灾难的
visible	[ˈvizəbl]	adj.看得见的，明显的，显著的 n.可见物
transcend	[trænˈsend]	vt.超越，胜过
boundary	[ˈbaundəri]	n.边界，分界线
omnipotent	[ɔmˈnipətənt]	adj.全能的，无所不能的
visualization	[ˌvizjuəlaiˈzeiʃən]	n.可视化
span	[spæn]	n.跨度，跨距，范围
contender	[kənˈtendə]	n.竞争者
volatility	[ˌvɔləˈtiliti]	n.波动率；波动性；波动
variability	[ˌveəriəˈbiliti]	n.变异性；可变性
cornerstone	[ˈkɔ:nəstəun]	n.奠基石，基础，最重要的部分
viability	[ˌvaiəˈbiliti]	n.可行性，切实可行，能办到；生存能力
gigantic	[dʒaiˈgæntik]	adj.巨人般的，巨大的
proportion	[prəˈpɔ:ʃən]	n.比例；均衡；部分 vt.使成比例；使均衡，分摊
guzzle	[ˈgʌzl]	vt.狂饮，暴食；消耗
fascinate	[ˈfæsineit]	vt.使……着迷，使……神魂颠倒 vi.入迷，极度迷人的
fad	[fæd]	n.时尚，一时流行的狂热，一时的爱好
funnel	[ˈfʌnəl]	vt.& vi.把……灌进漏斗；使成漏斗状；成漏斗形；使汇集 n.漏斗；漏斗状物

futility	[fjuːˈtiləti]	n. 无益，无用
nutshell	[ˈnʌtʃel]	n. 简言之，一言以蔽之
exemption	[igˈzempʃən]	n. 解除，免除
vitality	[vaiˈtæliti]	n. 时效性；动态性，灵活
criticality	[kritiˈkæliti]	n. 临界点；临界状态；紧急程度，危险程度
prioritize	[praiˈɔritaiz]	vt. 把……区分优先次序
pragmatic	[prægˈmætik]	adj. 实际的，注重实效的
arbitrage	[ˈɑːbitridʒ]	n. 套汇，套利交易
correlate	[ˈkɔrileit]	vt. 使相互关联
		vi. 和……相关
incomplete	[ˌinkəmˈpliːt]	adj. 不完全的，不完善的

Phrases

big data	大数据
capitalize on	充分利用；资本化
holy grail	圣杯；无处寻觅的稀世珍宝，努力却无法得到的东西
extracted ... from	从……中抽取，从……中提取
data warehouse	数据仓库
business intelligence tool	商业智能工具
information explosion	信息爆炸，知识爆炸
be superseded by ...	被……取代
wake up	活跃起来；引起注意；（使）认识到
invade into	侵入
unstructured text	非结构化文本
underlying data	源数据；基础数据；基本数据
pertain to	属于，关于，附属
in the attempt to	试图，企图
at large	普遍的；一般的；整体的
according to	依照
in a snap	立刻，马上
standard deviation	标准差，标准偏差
scale up	按比例增加，按比例提高
get fascinated with	迷上，沉溺于
in a nutshell	简言之，一言以蔽之
analysis paralysis	过度分析
be replaced with	由……代替
be aligned to	与……一致

be derived from	来自，源于
draw from...	从……抽取
consist of	构成，组成
customer contact center	客户联络中心，客户服务中心

Abbreviations

IT (Information Technology)	信息技术
IoT (Internet of Things)	物联网

Notes

[1] One has to demonstrate value that can be extracted from big or small data in order to justify the investments, whether on big data or on traditional analytics, data warehouse or business intelligence tools, whatever may be the buzzing nomenclature.

本句中，that can be extracted from big or small data 是一个定语从句，修饰和限定 value。in order to justify the investments, whether on big data or on traditional analytics, data warehouse or business intelligence tools 是一个目的状语从句，修饰主句的谓语 demonstrate。whatever may be the buzzing nomenclature 是一个让步状语从句。

[2] In the context of business operations that have not yet been touched by social media or IoT, the velocity arises from sophisticated enterprise applications that capture each and every minute detail involved in the completion of a particular business process.

本句中，that have not yet been touched by social media or IoT 是一个定语从句，修饰和限定 business operations。that capture each and every minute detail involved in the completion of a particular business process 也是一个定语从句，修饰和限定 enterprise applications。在该从句中，involved in the completion of a particular business process 是一个过去分词短语，做后置定语，修饰和限定 each and every minute detail。

[3] Since we are living in an increasingly digital world where technology has invaded into our glasses and watches, the variety of data that is generated is mind-boggling.

本句中，Since we are living in an increasingly digital world where technology has invaded into our glasses and watches 是一个原因状语从句，修饰和限定主句的谓语 is mind-boggling。在该从句中，where technology has invaded into our glasses and watches 也是一个定语从句，修饰和限定 digital world。that is generated 是一个定语从句，修饰和限定 the variety of data。

[4] Embedded in the concept of value is the need to check the viability of any project.

本句是一个表语前置的倒装句。the need to check the viability of any project 是主语，Embedded in the concept of value 是表语。正常语序应为：The need to check the viability of any project is embedded in the concept of value.

✎ Exercises

【Ex. 1】 根据课文内容回答问题。

1. What can insights from big data do?
2. What does velocity refer to? What are some of the factors that exacerbate this trend?
3. Why is the variety of data that is generated is mind-boggling?
4. What does veracity imply?
5. What have always existed within enterprises and have been one of the major roadblocks in the attempt to extract value from data?
6. What should relevant information be?
7. How many hours of video are uploaded every minute on YouTube according to official statistics?
8. What is one of the cornerstones of traditional statistics?
9. What kind of information needs to be prioritized?
10. Where is the word vincularity derived from? What does it imply?

【Ex. 2】 把下列句子翻译为中文。

1. I hope that this talk has given you some insight into the kind of the work that we've been doing.
2. The new systems have been optimized for running Microsoft Windows.
3. These designs demonstrate her unerring eye for colour and detail.
4. Let me make this clear: A bar chart is not analytics.
5. A good dictionary will give us the connotation of a word as well as its denotation.
6. The latest lifestyle trend is downshifting.
7. The end of an era presupposes the start of another.
8. You cannot combine structured and unstructured exception handling in the same function.
9. Finally, the practical application shows the feasibility and veracity of this approach.
10. The viability of multilayer switches depends on the protocol supported.

【Ex. 3】 短文翻译。

Cloud computing is a general term for anything that involves delivering hosted services over the Internet. These services are broadly divided into three categories: Infrastructure-as-a-Service (IaaS), Platform-as-a-Service (PaaS) and Software-as-a-Service (SaaS). The name cloud computing was inspired by the cloud symbol that's often used to represent the Internet in flowcharts and diagrams.

A cloud service has three distinct characteristics that differentiate it from traditional hosting. It is sold on demand, typically by the minute or the hour; it is elastic — a user can have as much or as little of a service as they want at any given time; and the service is fully managed by the provider (the consumer needs nothing but a personal computer and Internet access). Significant innovations in virtualization and distributed computing, as well as improved access to high-speed Internet and a weak economy, have accelerated interest in cloud computing.

【Ex. 4】 将下列词填入适当的位置（每词只用一次）。

| update | media | reduced | meaningful | challenge |
| multiple | data | recent | messages | explosion |

Volume

We currently see the exponential growth in the data storage as the data is now more than text data. We can find data in the format of videos, musics and large images on our social __(1)__ channels. It is very common to have Terabytes and Petabytes of the storage system for enterprises. As the database grows the applications and architecture built to support the __(2)__ needs to be reevaluated quite often. Sometimes the same data is reevaluated with multiple angles and even though the original data is the same the new found intelligence creates __(3)__ of the data. The big volume indeed represents Big Data.

Velocity

The data growth and social media explosion have changed how we look at the data. There was a time when we used to believe that data of yesterday is __(4)__. The matter of the fact newspapers is still following that logic. However, news channels and radios have changed how fast we receive the news. Today, people reply on social media to __(5)__ them with the latest happening. On social media sometimes a few seconds old messages (a tweet, status updates etc.) is not something interests users. They often discard old __(6)__ and pay attention to recent updates. The data movement is now almost real time and the update window has __(7)__ to fractions of the seconds. This high velocity data represent Big Data.

Variety

Data can be stored in __(8)__ format. For example DataBase, Excel, CSV, ACCESS or for the matter of the fact, it can be stored in a simple text file. Sometimes the data is not even in the traditional format as we assume, it may be in the form of video, SMS, pdf or something we might have not thought about it. It is the need of the organization to arrange it and make it __(9)__. It will be easy to do so if we have data in the same format, however it is not the case most of the time. The real world have data in many different formats and that is the __(10)__ we need to overcome with the Big Data. This variety of the data represent Big Data.

Text B

Big Data Analytics

Big data analytics is the process of collecting, organizing and analyzing large sets of data (called big data) to discover patterns and other useful information. Big data analytics can help organizations to better understand the information contained within the data and will also help identify the data that is most important to the business and future business decisions. Analysts working with big data basically want the knowledge that comes from analyzing the data.

1. Big Data Requires High Performance Analytics

To analyze such a large volume of data, big data analytics is typically performed using specialized software tools and applications for predictive analytics, data mining, text mining, forecasting and data optimization. Collectively these processes are separate but highly integrated functions of high-performance analytics. Using big data tools and software enables an organization to process extremely large volumes of data that a business has collected to determine which data is relevant and can be analyzed to drive better business decisions in the future.

2. The Challenges of Big Data Analytics

For most organizations, big data analysis is a challenge. Consider the sheer volume of data and the different formats of the data (both structured and unstructured data) that is collected across the entire organization and the many different ways different types of data can be combined, contrasted and analyzed to find patterns and other useful business information.

The first challenge is in breaking down data silos to access all data an organization stores in different places and often in different systems. A second big data challenge is in creating platforms that can pull in unstructured data as easily as structured data. This massive volume of data is typically so large that it's difficult to process using traditional database and software methods.

3. How Big Data Analytics is Used Today

As the technology that helps an organization to break down data silos and analyze data improves, business can be transformed in all sorts of ways. According to Datamation, today's advances in analyzing big data allow researchers to decode human DNA in minutes, predict where terrorists plan to attack, determine which gene is mostly likely to be responsible for certain diseases and, of course, which ads you are most likely to respond to on Facebook.

Another example comes from one of the biggest mobile carriers in the world. France's Orange launched its Data for Development project by releasing subscriber data for customers in the Ivory Coast. The 2.5 billion records, which were made anonymous, included details on calls and text messages exchanged between 5 million users. Researchers accessed the data and sent Orange proposals for how the data could serve as the foundation for development projects to improve public health and safety. Proposed projects included one that showed how to improve public safety by tracking cell phone data to map where people went after emergencies; another showed how to use cellular data for disease containment.

4. The Benefits of Big Data Analytics

Enterprises are increasingly looking to find actionable insights into their data. Many big data projects originate from the need to answer specific business questions. With the right big data analytics platforms in place, an enterprise can boost sales, increase efficiency, and improve operations, customer service and risk management.

Webopedia parent company, QuinStreet, surveyed 540 enterprise decision-makers involved in big data purchases to learn which business areas companies plan to use Big Data analytics to improve operations. About half of all respondents said they were applying big data analytics to improve customer retention, help with product development and gain a competitive advantage.

Notably, the business area getting the most attention relates to increasing efficiency and optimizing operations. Specifically, 62 percent of respondents said that they use big data analytics to improve speed and reduce complexity.

5. Top 10 Hot Big Data Technologies

As the big data analytics market rapidly expands to include mainstream customers, which technologies are most in demand and promise the most growth potential? The answers can be found in TechRadar: Big Data, Q1 2016, a new Forrester Research report evaluating the maturity and trajectory of 22 technologies across the entire data life cycle. The winners all contribute to real-time, predictive, and integrated insights, what big data customers want now.

Here is my talk on the 10 hottest big data technologies based on Forrester's analysis:

5.1 Predictive analytics

Software and/or hardware solutions that allow firms to discover, evaluate, optimize, and deploy predictive models by analyzing big data sources to improve business performance or mitigate risk.

5.2 NoSQL databases

Key-value, document, and graph databases.

5.3 Search and knowledge discovery

Tools and technologies to support self-service extraction of information and new insights from large repositories of unstructured and structured data that resides in multiple sources such as file systems, databases, streams, APIs, and other platforms and applications.

5.4 Stream analytics

Software that can filter, aggregate, enrich, and analyze a high throughput of data from multiple disparate live data sources and in any data format.

5.5 In-memory data fabric

Provides low-latency access and processing of large quantities of data by distributing data across the dynamic random access memory (DRAM), Flash, or SSD of a distributed computer system.

5.6 Distributed file stores

A computer network where data is stored on more than one node, often in a replicated fashion, for redundancy and performance.

5.7 Data virtualization

A technology that delivers information from various data sources, including big data sources such as Hadoop and distributed data stores in real-time and near-real time.

5.8 Data integration

Tools for data orchestration across solutions such as Amazon Elastic MapReduce (EMR), Apache Hive, Apache Pig, Apache Spark, MapReduce, Couchbase, Hadoop, and MongoDB.

5.9 Data preparation

Software that eases the burden of sourcing, shaping, cleansing, and sharing diverse and messy data sets to accelerate data's usefulness for analytics.

5.10 Data quality

Products that conduct data cleansing and enrichment on large, high-velocity data sets, using parallel operations on distributed data stores and databases.

New Words

analytic	[ˌænəˈlitik]	*adj.*分析的，解析的
predictive	[priˈdiktiv]	*adj.*预言性的，成为前兆的
forecasting	[ˈfɔːkɑːstiŋ]	*n.*预测
collectively	[kəˈlektivli]	*adv.*全体地，共同地
sheer	[ʃiə]	*adj.*全然的，纯粹的，绝对的
combine	[kəmˈbain]	*vt.*组合，结合
contrast	[ˈkɔntræst]	*vt.*使与……对比，使与……对照 *vi.*和……形成对照 *n.*对比，对照，(对照中的)差异
silo	[ˈsailəu]	*n.*竖井
datamation	[ˌdeitəˈmeiʃən]	*n.*自动化资料处理
researcher	[riˈsəːtʃə]	*n.*研究者
predict	[priˈdikt]	*v.*预知，预言，预报
terrorist	[ˈterərist]	*n.*恐怖分子
gene	[dʒiːn]	*n.*[遗传]基因

disease	[di'zi:z]		n.疾病，弊病
anonymous	[ə'nɔniməs]		adj.匿名的
emergency	[i'mə:dʒnsi]		n.紧急情况，突然事件，非常时刻，紧急事件
containment	[kən'teinmənt]		n.控制，遏制政策
boost	[bu:st]		v.推进
retention	[ri'tenʃən]		n.保持力
mainstream	['meinstri:m]		n.主流
maturity	[mə'tjuəriti]		n.成熟，完备
trajectory	['trædʒiktəri]		n.轨道，弹道
mitigate	['mitigeit]		v.减轻
self-service	['self-'sə:vis]		n.自助式
enrich	[in'ritʃ]		vt.浓缩
low-latency	[ləu-'leitənsi]		n.低反应期，短反应时间
orchestration	[ˌɔ:ki'streiʃən]		n.管弦乐编曲
burden	['bə:dən]		n.担子，负担 v.负担
messy	['mesi]		adj.凌乱的，杂乱
usefulness	['ju:sfulnis]		n.有用，有效性

Phrases

high performance	高性能，高精确度
text mining	文本挖掘
in breaking down	在打破……
data silo	数据竖井，数据孤岛
subscribe for	预订，认购
text message	短信，短消息
cell phone	手机
disease containment	疾病控制
originate from	发源于
customer service	客户服务
competitive advantage	竞争优势
life cycle	生命周期
business performance	经营成绩，经营业绩
multiple source	多个来源，复合源
distributed computer system	分布式计算机系统
parallel operation	平行工作

Abbreviations

DNA (Deoxyribonucleic Acid)　　脱氧核糖核酸
SSD (Solid State Drives)　　固态硬盘

Exercises

【Ex. 5】根据课文内容回答问题。
(1) What is big data analytics?
(2) What can big data analytics do?
(3) How is big data analytics typically performed to analyze such a large volume of data?
(4) What is the first challenge of big data analytics?
(5) What is a second big data challenge?
(6) What do today's advances in analyzing big data allow researchers to do according to Datamation?
(7) What do many big data projects originate from?
(8) What can an enterprise do with the right big data analytics platforms in place?
(9) What does the business area getting the most attention relate to?
(10) What is the last part of the passage mainly talk about?

参考译文

大 数 据

大数据正在改变组织内部人们协同工作的方式。它正在创造一种文化，使得业务和 IT 领导者必须联合起来，以便实现所有数据的价值。大数据让所有员工能够更好地做出决策——包括深化客户参与度、优化运营、防止威胁和欺诈行为以及开辟新的收入来源。

1. 大 V

1.1 价值

这的确是大数据的梦想，也是我们在寻找的目标。从大大小小的数据中获得价值以证明投资所值，无论是大数据分析或传统分析、数据仓库或商业智能工具，或许只是不同的名称而已。根据谷歌搜索过去两年寻找类似条目的数量，似乎表明人们对大数据的价值越来越感兴趣。

1.2 数据量

毫无疑问，信息爆炸已经重新定义了数据量的含义。有几个如此惊人的统计数字，要跟踪数据越来越难了，要度量这样的数据需要给"字节"前面加上种种前缀。因为有"巨量的数据"，新创造出的术语"Hellabyte"已经超越 PB、EB、ZB 和 YB。然而，这些度量单位将被 Brontobytes、Geopbyte 等替代，让我们继续吧！

1.3 高速性

同样地，高速性是指产生数据的速度。社交媒体的扩散和 IoT（物联网）的爆炸式增长是加剧这一趋势的一些因素。在尚未被社交媒体或物联网影响的业务运营中，时效性来自复杂的企业应用，它捕捉了每一个特定业务流程的每一个微小的细节。传统上企业应用也捕获这些信息，但在大数据时代，这些信息就是力量。

1.4 多样性

大数据的最后一个原始属性是多样性。既然我们生活在一个日益数字化的世界里，技术已经侵入我们的眼镜和手表，多样性所产生的数据是令人难以置信的。可用的计算能力能够处理非结构化的文本、图像、音频、视频以及来自物联网传感器的数据，这几乎可以捕获我们周围的一切。今天，大数据的这个属性与我们现在的生活的联系比以往更紧密。

1.5 真实性或有效性

数据的真实性或有效性对提取基础数据的价值非常重要。真实性意味着数据是可验证的和真实的。如果违反这个条件，其结果可能是灾难性的。更重要的是，有几种情况，其中数据虽然准确但在特定情况下无效。例如，如果试图确定谷歌中"大数据"的搜索量，我们也会获得有关"大数据"的"危险"的结果。

1.6 可见性

信息孤岛一直在企业中存在，并且一直是从数据中提取价值的主要障碍之一。不仅应该有相关信息，而且应该在合适的时间给合适的人看到。可操作的数据需要超越职能部门甚至组织的界限，并被其所见，才能释放数据的价值。个体可能会认为在他们手中的信息就是力量，但在大数据时代，大量的对全球有效的整合信息才真正无所不能！

1.7 视觉性

我们生活在一个日益视觉化的世界，统计数据表明，在互联网上共享的图像和视频的数量以惊人的速度增加。据官方统计，每分钟有 300 小时的视频被上传到 YouTube。在商业环境中，适当的可视化数据对管理者是至关重要的，他们能够在有限的时间和资源甚至更有限的注意力中获得价值！

2. 更多的属性

除了上述的 7 V，可能还有其他几个 V。

2.1 波动率

随着越来越多的应用（如 SnapChat 和物联网传感器）出现，可能即时产生一些输入和输出数据。基础数据源的波动率将来可能成为其定义属性之一。

2.2 变异性

传统统计的一个基石是标准差和变异。无论它在不在大数据的扩展列表中，都绝不能被忽略。

2.3 可行性

每个项目的可行性都需要检查，这包含在价值概念之中。大数据项目可占据巨大的比例并非常快地消耗大量资源。谁不快速学习并沉迷时尚之中，就会耗尽资金而失败。简言之，任何项目都要进行可行性研究，大数据项目也不例外，无论它是否仍然是一个流行词。

2.4 时效性

数据的时效性或关键性是另一至关重要的概念，它包含在价值概念之中。应该优先考虑对实现基础商业目标更有意义或更重要的信息。需要用更务实的方法来取代过度分析。技术允许营销人员创建一个片段，但这样极端的分割对组织至关重要吗？它与组织战略一致吗？

2.5 连通性

Vincularity 这个词汇源于拉丁语，意思是连通性或链接。这个概念与当今的互联世界密切相关。连接不同信息集合可以得到潜在的套利价值。例如，政府一直尝试把主要支出的细节相连接，并将其与收入报税单相关联以发现是否隐瞒收入。而这一目的，现在可以通过从社交媒体的帖子上提取信息来实现。

3. 一个大数据的示例

大数据的一个例子可能是 PB 级数据（1024 兆兆字节）或艾字节（1024 千兆兆字节），它包含了数百万人数十亿的记录——来自不同信息源（如网络、销售、客户联络中心、社交媒体及移动数据等）。该数据通常结构性不强，而且往往是不完整的和难以访问的。

Unit 2

Text A

Computer Software

Computer software, consisting of programs, enables a computer to perform specific tasks. It is opposed to its physical components (hardware) which can only do the tasks they are mechanically designed for. The term includes application software such as word processors, which perform productive tasks for users, system software such as operating systems, which interface with hardware to run the necessary services for user-interfaces and applications, and middleware, which controls and coordinates distributed systems.

1. Terminology

The term "software" is an instruction-procedural programming source for scheduling instruction streams according to the von Neumann machine paradigm. It should not be confused with Configware and Flowware, which are programming sources for configuring the resources (structural "programming" by Configware) and for scheduling the data streams (data-procedural programming by Flowware) of the Anti machine paradigm of Reconfigurable Computing systems.

2. Relationship to Computer Hardware

Computer software is so called in contrast to computer hardware, which encompasses the physical interconnections and devices required to store and execute (or run) the software. In computers, software is loaded into RAM and executed in the central processing unit. At the lowest level, software consists of a machine language specific to an individual processor. A machine language consists of groups of binary values signifying processor instructions (object codes), which change the state of the computer from its preceding state. Software is an ordered sequence of instructions for changing the state of the computer hardware in a particular sequence. It is usually written in high-level programming languages that are easier and more efficient for humans to use (closer to natural language) than machine language. High-level languages are compiled or interpreted into machine language object code. Software may also be written in an assembly language, essentially, a mnemonic representation of a machine language using a natural language alphabet. Assembly language must be assembled into object code via an assembler.

In computer science and software engineering, computer software is all computer programs. The concept of reading different sequences of instructions into the memory of a device to control computations was invented by Charles Babbage as part of his difference engine.

3. Types

Practical computer systems divide software systems into three major classes: system software, programming software and application software, although the distinction is arbitrary, and often blurred.

3.1 System Software

System software helps run the computer hardware and computer system. It includes operating systems, device drivers, diagnostic tools, servers, windowing systems, utilities and more. The purpose of systems software is to insulate the applications programmer as much as possible from the details of the particular computer complex being used, especially memory and other hardware features, and such accessory devices as communications, printers, readers, displays, keyboards, etc.

3.2 Programming Software

Programming software usually provides tools to assist a programmer in writing computer programs and software using different programming languages in a more convenient way. The tools include text editors, compilers, interpreters, linkers, debuggers, and so on. An integrated development environment (IDE) merges those tools into a software bundle, and a programmer may not need to type multiple commands for compiling, interpreting, debugging, tracing, and etc., because the IDE usually has an advanced graphical user interface, or GUI.

3.3 Application Software

Application software allows end users to accomplish one or more specific (non-computer related) tasks. Typical applications include industrial automation, business software, educational software, medical software, databases, and computer games. Businesses are probably the biggest users of application software, but almost every field of human activity now uses some form of application software. It is used to automate all sorts of functions.

4. Three Layers

Users often see things differently than programmers. People who use modern general purpose computers (as opposed to embedded systems, analog computers, supercomputers, etc.) usually see three layers of software performing a variety of tasks: platform, application, and user software.

4.1 Platform Software

Platform includes the firmware, device drivers, an operating system, and typically a graphical user interface which, in total, allows a user to interact with the computer and its peripherals (associated equipment). Platform software often comes bundled with the computer, and users may not realize that it exists or that they have a choice to use different platform software.

4.2 Application Software

Application software or Applications are what most people think of when they think of software. Typical examples include office suites and video games. Application software is often purchased separately from computer hardware. Sometimes applications are bundled with the computer, but that does not change the fact that they run as independent applications. Applications are almost always independent programs from the operating system, though they are often tailored for specific platforms. Most users think of compilers, databases, and other "system software" as applications.

4.3 User Software

User software tailors systems to meet the users specific needs. User software include spreadsheet templates, word processor macros, scientific simulations, and scripts for graphics and animations. Even email filters are a kind of user software. Users create this software themselves and often overlook how important it is. Depending on how competently the user-written software has been integrated into purchased application packages, many users may not be aware of the distinction between the purchased packages, and what has been added by fellow co-workers.

5. Operation

Computer software has to be "loaded" into the computer's storage (such as a hard drive, memory, or RAM). Once the software is loaded, the computer is able to execute the software. Computers operate by executing the computer program. This involves passing instructions from the application software, through the system software, to the hardware which ultimately receives the instruction as machine code. Each instruction causes the computer to carry out an operation — moving data, carrying out a computation, or altering the control flow of instructions.

Data movement is typically from one place in memory to another. Sometimes it involves moving data between memory and registers which enable high-speed data access in the CPU. Moving data, especially large amounts of it, can be costly. So, this is sometimes avoided by using "pointers" to data instead. Computations include simple operations such as incrementing the value of a variable data element. More complex computations may involve many operations and data elements together.

Instructions may be performed sequentially, conditionally, or iteratively. Sequential instructions are those operations that are performed one after another. Conditional instructions are performed such that different sets of instructions execute depending on the value(s) of some data. In some languages this is known as an "if" statement. Iterative instructions are performed repetitively and may depend on some data value. This is sometimes called a "loop." Often, one instruction may "call" another set of instructions that are defined in some other program or module. When more than one computer processor is used, instructions may be executed simultaneously.

A simple example of the way software operates is what happens when a user selects an entry such as "Copy" from a menu. In this case, a conditional instruction is executed to copy text from data in a "document" area residing in memory, perhaps to an intermediate storage area known as a "clipboard" data area. If a different menu entry such as "Paste" is chosen, the software may execute the instructions to copy the text from the clipboard data area to a specific location in the same or another document in memory.

Depending on the application, even the example above could become complicated. The field of software engineering endeavors to manage the complexity of how software operates. This is especially true for software that operates in the context of a large or powerful computer system.

Kinds of software by operation: computer program as executable, source code or script, configuration.

6. Quality and reliability

Software reliability considers the errors, faults, and failures related to the creation and operation of software.

Software quality is very important, especially for commercial and system software like Microsoft Office, Microsoft Windows and Linux. If software is faulty (buggy), it can delete a person's work, crash the computer and do other unexpected things. Faults and errors are called "bugs" which are often discovered during alpha and beta testing. Software is often also a victim to what is known as software aging, the progressive performance degradation resulting from a combination of unseen bugs.

Many bugs are discovered and eliminated (debugged) through software testing. However, software testing rarely—if ever—eliminates every bug; some programmers say that "every program has at least one more bug". In the waterfall method of software development, separate testing teams are typically employed, but in newer approaches, collectively termed agile software development, developers often do all their own testing, and demonstrate the software to users/clients regularly to obtain feedback. Software can be tested through unit testing, regression testing and other methods, which are done manually, or most commonly, automatically, since the amount of code to be tested can be quite large. For instance, NASA has extremely rigorous software testing procedures for many operating systems and communication functions. Many NASA-based operations interact and identify each other through command programs. This enables many people who work at NASA to check and evaluate functional systems overall. Programs containing command software enable hardware engineering and system operations to function much easier together.

New Words

mechanically	[mi'kænikəli]	adv.机械地
middleware	['midlwɛə]	n.中间设备，中间件
procedural	[prə'si:dʒərəl]	adj.程序上的
paradigm	['pærədaim]	n.范例
structural	['strʌktʃərəl]	adj.结构的，结构化
interconnection	[ˌintəkə'nekʃən]	n.互相连接
compile	[kəm'pail]	vt.编译
assembler	[ə'semblə]	n.汇编程序
arbitrary	['ɑ:bitrəri]	adj.武断的，专断的
blur	[blə:]	v.模糊
insulate	['insjuleit]	vt.隔离，使绝缘

reader	[ˈriːdə]	n.读卡机
convenient	[kənˈviːnjənt]	adj.便利的，方便的
interpreter	[inˈtəːpritə]	n.解释程序
linker	[ˈliŋkə]	n.(目标代码)连接器
debugger	[diːˈbʌgə]	n.调试器
merge	[məːdʒ]	v.合并，并入，融合
bundle	[ˈbʌndl]	n.捆，束，包 v.捆扎
tailored	[ˈteiləd]	adj.定做的，特制的，专门的
template	[ˈtemplit]	n.模板(=templet)
macro	[ˈmækrəu]	n.宏
script	[skript]	n.脚本
animation	[ˌæniˈmeiʃən]	n.动画
filter	[ˈfiltə]	n.过滤器，滤波器
competently	[ˈkɔmpitəntli]	adv.胜任地，适合地
co-worker	[ˈkəuˈwəːkə]	n.合作者，同事，帮手
alter	[ˈɔːltə]	v.改变
costly	[ˈkɔstli]	adj.昂贵的，困难的；造成损失的
pointer	[ˈpɔintə]	n.指针
conditionally	[kənˈdiʃnəli]	adv.有条件地
iteratively	[ˈitərətivli]	adv.反复地；迭代地
call	[kɔːl]	n. & v.调用
clipboard	[ˈklipbɔːd]	n.剪贴板
endeavor	[inˈdevə]	n. & vi.尽力，努力
reliability	[riˌlaiəˈbiliti]	n.可靠性

🖊 Phrases

be opposed to	与……相对，和……相反
system software	系统软件
distributed system	分布式的计算机系统
be confused with	混淆
in contrast to	和……形成对比，和……形成对照
machine language	机器语言
object code	结果代码
ordered sequence	有序序列
high-level programming languages	高级编程语言

natural language	自然语言
assembly language	汇编语言
software engineering	软件工程
difference engine	差分机
divide … into …	把……分成……
device driver	设备驱动程序
diagnostic tool	诊断工具
as much as possible	尽可能
computer complex	计算装置
text editor	文本编辑器
integrated development environment (IDE)	集成开发环境
computer game	计算机游戏程序
all sorts of	各种各样的
embedded system	嵌入式系统
analog computer	模拟计算机
a variety of	多种的
platform software	平台软件
in total	整个地(=as a whole)
come with	伴随……发生，与……一起供给
video game	计算机视频游戏，电视游戏
separate from	分离，分开
be integrated into …	统一到……中，整合到……中
be aware of	知道
be able to	能，会
carry out	完成，实现，执行
one after another	接连地
be known as	被认为是
conditional instruction	条件指令
be incapable of	不能
source code	源编码，源代码，源程序
software reliability	软件可靠性

Notes

[1] It is opposed to its physical components (hardware) which can only do the tasks they are mechanically designed for.

本句中，which can only do the tasks they are mechanically designed for 是一个定语从句，

修饰和限定 its physical components。在该从句中，they are mechanically designed for 也是一个定语从句，修饰和限定 the tasks。(hardware)是对 its physical components 的补充说明。

[2] The term includes application software such as word processors, which perform productive tasks for users, system software such as operating systems, which interface with hardware to run the necessary services for user-interfaces and applications, and middleware, which controls and coordinates distributed systems.

本句中，which perform productive tasks for users 是一个非限定性定语从句，修饰 word processors；which interface with hardware to run the necessary services for user-interfaces and applications 是一个非限定性定语从句，修饰 operating systems；which controls and coordinates distributed systems 也是一个非限定性定语从句，修饰 middleware。such as 的意思是"例如"，用来举例说明。

[3] It should not be confused with Configware and Flowware, which are programming sources for configuring the resources (structural "programming" by Configware) and for scheduling the data streams (data-procedural programming by Flowware) of the Anti machine paradigm of Reconfigurable Computing systems.

本句中，which are programming sources for configuring the resources (structural "programming" by Configware) and for scheduling the data streams (data-procedural programming by Flowware) of the Anti machine paradigm of Reconfigurable Computing systems 是一个非限定性定语从句，对 Configware and Flowware 进行补充说明。

[4] Computer software is so called in contrast to computer hardware, which encompasses the physical interconnections and devices required to store and execute (or run) the software.

本句中，which encompasses the physical interconnections and devices required to store and execute (or run) the software 是一个非限定性定语从句，对 computer hardware 进行补充说明。required to store and execute (or run) the software 是一个过去分词短语，作定语，修饰和限定 the physical interconnections and devices。in contrast to 的意思是"与……形成对比"，"相比之下"。

[5] A machine language consists of groups of binary values signifying processor instructions (object codes), which change the state of the computer from its preceding state.

本句中，signifying processor instructions (object code)是一个现在分词短语，作定语，修饰和限定 binary values。which change the state of the computer from its preceding state 是一个非限定性定语从句，对 processor instructions 进行补充说明。

Exercises

【Ex. 1】根据课文内容，回答以下问题。

(1) What does computer software consist of? What does it do?

(2) What does the term computer software include?
(3) What does a machine language consist of?
(4) What are the three major classes practical computer systems divide software systems into?
(5) What does system software do? What does it include?
(6) What does programming software usually do?
(7) What does application software usually do? What do typical applications include?
(8) What are the three layers of software?
(9) What are the sequential instructions, conditional instructions and iterative instructions respectively?
(10) What can happen if software is faulty (buggy)?

【Ex. 2】英汉互译

1.	device driver	1.	
2.	middleware	2.	
3.	diagnostic tool	3.	
4.	Integrated development environment	4.	
5.	pointer	5.	
6.	配置，设定	6.	
7.	嵌入式系统	7.	
8.	模块	8.	
9.	汇编程序	9.	
10.	编译器	10.	

【Ex. 3】将下列词填入适当的位置（每词只用一次）。

application	computer	create	enable	transferring
programs	software	user	basics	processing

System software is closely related to, but distinct from Operating System software. It is any computer software that provides the infrastructure over which __(1)__ can operate, i.e. it manages and controls computer hardware so that __(2)__ software can perform. Operating systems, such as GNU, Microsoft Windows, Mac OS X or Linux, are prominent examples of system __(3)__.

System software is software that basically allows the parts of a __(4)__ to work together. Without the system software the computer cannot operate as a single unit. In contrast to system software, software that allows you to do things like __(5)__ text documents, play games, listen to music, or surf the web.

In general, application programs are software that __(6)__ the end-user to perform specific, productive tasks, such as word __(7)__ or image manipulation. System software performs tasks like __(8)__ data from memory to disk, or rendering text onto a display device.

System software is not generally what a user would buy a computer for, instead, it is usually the __(9)__ of a computer which come built-in. Application software is the programs on the computer when the __(10)__ buys it. These may include word processors and web browsers.

【Ex. 4】把下列短文翻译成中文。

Computer programs (also software programs, or just programs) are instructions for a computer. A computer requires programs to function, typically executing the program's instructions in a central processor. The program has an executable form that the computer can use directly to execute the instructions.

Computer source code is often written by professional computer programmers. Source code may be converted into an executable file (sometimes called an executable program or a binary) by a compiler. Alternatively, computer programs may be executed by a central processing unit with the aid of an interpreter, or may be embedded directly into hardware.

Computer programs may be categorized along functional lines: system software and application software. And many computer programs may run simultaneously on a single computer, a process known as multitasking.

Text B

Software Development Process

Software development process is a structure imposed on the development of a software product. Synonyms include software life cycle and software process. There are several models for such processes, each describing approaches to a variety of tasks or activities that take place during the process.

1. Processes and meta-processes

A growing body of software development organizations implement process methodologies. Many of them are in the defense industry, which in the U.S. requires a rating based on 'process models' to obtain contracts. The international standard for describing the method of selecting, implementing and monitoring the life cycle for software is ISO 12207.

The Capability Maturity Model (CMM) is one of the leading models. Independent assessments grade organizations on how well they follow their defined processes, not on the quality of those processes or the software produced. CMM is gradually replaced by CMMI. ISO 9000 describes standards for formally organizing processes with documentation.

ISO 15504, also known as Software Process Improvement Capability Determination (SPICE), is a "framework for the assessment of software processes". This standard is aimed at setting out a clear model for process comparison. SPICE is used much like CMM and CMMI. It models processes to manage, control, guide and monitor software development. This model is then used to measure what a development organization or project team actually does during software development. This information is analyzed to identify weaknesses and drive improvement. It also identifies strengths that can be continued or integrated into common practice for that organization or team.

Six Sigma is a methodology to manage process variations that uses data and statistical analysis to measure and improve a company's operational performance. It works by identifying and eliminating defects in manufacturing and service-related processes. The maximum permissible defects is 3.4 per one million opportunities. However, Six Sigma is manufacturing-oriented and needs further research on its relevance to software development.

1.1 Domain Analysis

Often the first step in attempting to design a new piece of software, whether it be an addition to an existing software, a new application, a new subsystem or a whole new system, is, what is generally referred to as "Domain Analysis". Assuming that the developers (including the analysts) are not sufficiently knowledgeable in the subject area of the new software, the first task is to investigate the so-called "domain" of the software. The more knowledgeable they are about the domain already, the less the work required. Another objective of this work is to make the analysts who will later try to elicit and gather the requirements from the area experts or professionals, speak with them in the domain's own terminology and to better understand what is being said by these people. Otherwise they will not be taken seriously. So, this phase is an important prelude to extracting and gathering the requirements. The following quote captures the kind of situation an analyst who hasn't done his homework well may face in speaking with a professional from the domain: "I know you believe you understood what you think I said, but I am not sure you realize what you heard is not what I meant."

1.2 Software Elements Analysis

The most important task in creating a software product is extracting the requirements. Customers typically know what they want, but not what software should do, while incomplete, ambiguous or contradictory requirements are recognized by skilled and experienced software engineers. Frequently demonstrating live code may help reduce the risk that the requirements are incorrect.

1.3 Specification

Specification is the task of precisely describing the software to be written, possibly in a rigorous way. In practice, most successful specifications are written to understand and fine-tune applications that were already well-developed, although safety-critical software systems are often carefully specified prior to application development. Specifications are most important for external interfaces that must remain stable.

1.4 Software architecture

The architecture of a software system refers to an abstract representation of that system. Architecture is concerned with making sure the software system will meet the requirements of the product, as well as ensuring that future requirements can be addressed. The architecture step also addresses interfaces between the software system and other software products, as well as the underlying hardware or the host operating system.

1.5 Implementation (or coding)

Reducing a design to code may be the most obvious part of the software engineering job, but it is not necessarily the largest portion.

1.6 Testing

Testing of parts of software, especially where code by two different engineers must work together, falls to the software engineer.

1.7 Documentation

An important (and often overlooked) task is documenting the internal design of software for the purpose of future maintenance and enhancement. Documentation is most important for external interfaces.

2. Software Training and Support

A large percentage of software projects fail because the developers fail to realize that it doesn't matter how much time and planning a development team puts into creating software if nobody in an organization ends up using it. People are occasionally resistant to change and avoid venturing into an unfamiliar area so, as a part of the deployment phase, it is very important to have training classes for the most enthusiastic software users (build excitement and confidence), shifting the training towards the neutral users intermixed with the avid supporters, and finally incorporate the rest of the organization into adopting the new software. Users will have lots of questions and software problems which leads to the next phase of software.

3. Maintenance

Maintaining and enhancing software to cope with newly discovered problems or new requirements can take far more time than the initial development of the software. Not only may it be necessary to add code that does not fit the original design but just determining how software works at some point after it is completed may require significant effort by a software engineer. About 2/3 of all software engineering work is maintenance, but this statistic can be misleading. A small part of that is fixing bugs. Most maintenance is extending systems to do new things, which in many ways can be considered new work. In comparison, about 2/3 of all civil engineering, architecture, and construction work is maintenance in a similar way.

New Words

process	[prəˈses]	n. 过程；作用；方法，程序，步骤
		vt. 加工，处理
activity	[ækˈtiviti]	n. 活动，活动性；行动，行为
defense	[diˈfens]	n. 国防；防卫
contract	[ˈkɔntrækt]	n. 合同
assessment	[əˈsesmənt]	n. 估计，估算；评估，评价
grade	[greid]	n. 等级，级别
		vt. 评分，评级
gradually	[ˈgrædjuəli]	adv. 逐渐地
formally	[ˈfɔːməli]	adv. 正式地，形式上
team	[tiːm]	n. 队，组
actually	[ˈæktʃuəli]	adv. 实际上，事实上
weakness	[ˈwiːknis]	n. 弱点，缺点
methodology	[meθəˈdɔlədʒi]	n. 方法学，方法论
permissible	[pəˈmisibəl]	adj. 允许的，承认的
defect	[diˈfekt]	n. 过失，缺点
opportunity	[ˌɔpəˈtjuːniti]	n. 机会，时机

subsystem	[ˈsʌbˌsistim]	n.子系统
sufficiently	[səˈfiʃəntli]	adv.足够地，充分地
knowledgeable	[ˈnɔlidʒəbəl]	adj.博学的；有见识的
investigate	[inˈvestigeit]	v.调查，研究
elicit	[iˈlisit]	vt.得出，引出，抽出；引起
gather	[ˈgæðə]	n. & vi.集合，聚集
phase	[feiz]	n.阶段；状态
prelude	[ˈprelju:d]	n.先驱，前奏，序幕
ambiguous	[æmˈbigjuəs]	adj.不明确的
contradictory	[ˌkɔntrəˈdiktəri]	adj.矛盾的；反对的
skilled	[skild]	adj.熟练的
rigorous	[ˈrigərəs]	adj.严格的；精确的，一丝不苟的
fine-tune	[ˈfain-ˈtju:n]	v.调整；使有规则
obvious	[ˈɔbviəs]	adj.明显的，显而易见的
documentation	[ˌdɔkjumenˈteiʃən]	n.文件
overlook	[ˌəuvəˈluk]	vt.没注意到
enhancement	[inˈhɑ:nsmənt]	n.增强，促进
occasionally	[əˈkeiʒənəli]	adv.有时候，偶尔
resistant	[riˈzistənt]	adj.抵抗的，反抗的
venture	[ˈventʃə]	n.冒险；投机；风险 v.冒险
unfamiliar	[ˌʌnfəˈmiljə]	adj.新奇的；不熟悉的，没有经验的
deployment	[diˈplɔimənt]	n.部署
enthusiastic	[inˌθju:ziˈæstik]	adj.热心的，热情的
excitement	[ikˈsaitmənt]	n.刺激；兴奋，激动
confidence	[ˈkɔnfidəns]	n.信心
avid	[ˈævid]	adj.渴望的
incorporate	[ˈinkɔpərit]	vi.合并；混合
misleading	[misˈli:diŋ]	adj.易误解的，令人误解的
bug	[bʌg]	n.故障，问题

🕮 Phrases

impose on	利用；施加影响于
software life cycle	软件生命期

Capability Maturity Model (CMM)	（软件）能力成熟度模型
Software Process Improvement Capability Determination (SPICE)	软件过程改进能力测定
set out	表明；展示
Six Sigma	六西格玛
Domain Analysis	定义域分析
be concerned with	牵涉到，与……有关，参与
for the purpose of	为了，因……起见
put into	投入；使进入
end up	结束
cope with	处理；应付

✎ Abbreviations

CMMI (Capability Maturity Model Integration)　　能力成熟度集成模型

✎ Exercises

【Ex. 5】根据文章所提供的信息填空。

1. Software development process is a structure imposed on _____.
2. ISO 15504, also known as _____, is a "framework for the assessment of software processes". This standard is aimed at _____.
3. Six Sigma is a methodology _____ that uses data and statistical analysis _____.
4. Often the first step in attempting to design a new piece of software is _____.
5. The most important task in creating a software product is _____.
6. Specification is the task of _____.
7. Architecture is concerned with _____, as well as ensuring _____.
8. Testing of parts of software, especially where code by two different engineers must work together, falls to _____.
9. A large percentage of software projects fail because the developers fail to realize _____.
10. Most maintenance is extending systems to do new things, which in many ways can be considered _____.

参考译文

计算机软件

计算机软件由程序组成，可以让计算机执行特定的任务。它与只能机械地执行设定任务的物理构件（硬件）相对。这个术语包括应用程序（如能够提高用户工作效率的字处理器）、系统软件（如操作系统，它带有硬件接口，以便为用户界面和应用程序提供必需的服务）和中间件（管理与适应分布系统）。

1. 术语

术语"软件"是一个指令序列的程序源，它按照冯·诺依曼机制制定指令流，不应该把它与配置件和流件混淆。配置件和流件都是用来配置资源的程序源（通过配置件实现结构化"编程"），制定数据流（使用流件实现数据流程编程），是重配置计算机系统的反冯·诺依曼机制的范例。

2. 与计算机硬件的关系

计算机软件是与计算机硬件相对的称谓，硬件包括物理连接和存储与执行（或运行）软件所需的设备。在计算机中，软件装入 RAM 并在中央处理器中执行。最基本的软件可以由特定处理器的机器语言组成。机器语言由一组表示处理器指令（目标代码）的二进制值组成，这些目标代码可以改变计算机的状态。软件是有序的指令序列，以特定序列改变计算机硬件的状态。它通常用高级语言编写，对人来说比机器语言更便于理解且更有效（更接近自然语言）。高级语言可以编译或解释成机器语言目标代码。软件也可以用汇编语言编写，汇编语言本质上是用自然语言字母表示的机器语言助记形式。汇编语言必须通过编译器编译为目标代码。

在计算机科学和软件工程中，所有的计算机程序都是计算机软件。把不同的指令序列读到设备的内存以控制计算这一概念是由查尔斯·巴贝奇提出的，这成为其差分机的一部分。

3. 类型

实际的计算机系统把软件分为三大类：系统软件、编程软件和应用软件，尽管其差别是武断的，并且经常混淆。

3.1 系统软件

系统软件帮助运行计算机硬件和计算机系统。它包括操作系统、设备驱动程序、诊断工具、服务程序、窗口系统、实用程序等多种。系统软件的目的是把应用程序员与所用的复杂计算机的细节尽可能隔离开来，尤其是与内存和其他硬件、附件（如通信设备、打印机、阅读设备、显示器、键盘等）隔开。

3.2 编程软件

编程软件通常提供帮助程序员用不同的编程语言更方便地编写计算机程序和软件的工具。这些工具包括文本编辑器、编译器、解释程序、连接程序、调试程序等。集成开发环境把这些工具合并为一个软件包，程序员不用给编译、解释、调试、跟踪等操作输入多个命令，因为IDE通常有高级的图形用户界面或GUI。

3.3 应用软件

应用软件允许终端用户实现一个或多个（与计算机无关的）特定任务。典型的应用包括工业自动控制、商业软件、教育软件、医学软件、数据库和计算机游戏。商业大概是应用软件的最大用户，但几乎人类活动的每个领域现在都在使用某种应用软件。它用于各种各样的自动操作。

4. 三层

用户看待事情的方法往往与程序员不同。使用现代化普通计算机（与嵌入式计算机、模拟计算机、超级计算机等不同）的人往往认为执行各种操作的软件有三个层次：平台软件、应用软件和用户软件。

4.1 平台软件

平台软件包括固件、设备驱动程序、操作系统以及有代表性的图形用户界面。总体上说，图形用户界面让用户与计算机及外设（相关设备）交互。平台软件通常与计算机捆绑提供，用户可能没有意识到它的存在或者不知道他们可以选择其他平台软件。

4.2 应用软件

应用软件或应用就是大多数人认为的软件。典型的例子包括办公套件和视频游戏。应用软件通常与计算机硬件分开购买。有时应用软件也与计算机捆绑，但这不能改变它们作为独立应用软件而运行的事实。应用软件几乎总是独立于操作系统的程序，尽管它们通常为特定的平台而制作。大部分用户把编译程序、数据库和其他"系统软件"当作应用软件。

4.3 用户软件

用户软件定制多个系统以便满足用户的特定需求。用户软件包括电子表格模板、字处理程序的宏、科学仿真及用于图形和动画的脚本。甚至电子邮件过滤器也是用户软件的一种。用户自己建立用户软件，且通常忽视它的重要性。由于用户编写软件根据其适应性被整合到所购买的应用软件包中，因而许多用户不知道所购买的软件包的差别，也不知道合作伙伴在里面加了什么。

5. 运行

计算机软件必须被"装载"到计算机的存储器（如硬盘、存储器或 RAM）中。一旦软件被装入，计算机就可以执行该软件。计算机通过执行程序来运行。这包括从应用软件提取指令、经过系统软件发给最终以机器代码接收指令的硬件。每个指令都使计算机执行一个操作——移动数据、执行计算或改变指令的控制流。

数据移动通常是数据从内存中的一个位置向另一位置移动。有时数据也在内存和寄存器之间移动，寄存器可以实现在 CPU 中高速访问数据。移动数据——特别是移动大量的数据——是花费成本的。所以，有时使用"指针"来代替数据。计算包括简单的运算，如增加一个可变数据元素的值。更复杂的计算也许涉及许多运算和数据元素。

指令可以被连续地、有条件地或循环地执行。连续指令是一个接一个执行的操作。条件指令是根据某些数据的值执行不同的指令集合。在某些语言中，叫作 if 语句。循环指令是根据某些数值并反复地执行。这有时叫作一个"循环"。通常，一个指令可以调用另一个在其他程序或模块中定义的指令集合。当使用多个处理器时，指令可以同步执行。

这种软件运行方式的一个简单例子是，用户从一个菜单中选择一个菜单项（如 Copy）后所发生的一切。在这种情况下，条件指令被执行以便从内存驻留的文本区域的数据中复制一个文本到叫作"剪切板"的一个临时存储区域。如果另一菜单项（如 Paste）被选择，软件可以执行该指令，把"剪切板"数据区域中的文本复制到内存中同一文本或不同文本的特定位置。

根据应用软件，以上这个例子也可以变得复杂。软件工程致力于管理软件运行的复杂性。对于运行在大的或功能强的计算机系统的软件而言，尤其如此。

按照运行软件分为以下几种：可运行的计算机程序、源代码或脚本、配置程序。

6. 软件的质量和可靠性

软件可靠性考虑与软件建立和运行相关的错误、故障及失效。

软件质量非常重要，尤其是像 Microsoft Office、Microsoft Windows 和 Linux 这样的商业和系统软件。如果软件出现故障（出错），它可以删除一个人的工作，使计算机崩溃和做出其他意想不到的事情。故障和错误被称为"bug（漏洞）"，这是 alpha 和 beta

测试过程中经常出现的。软件通常也是所谓的软件老化的受害者，这源于看不见的错误组合而产生渐进的性能下降。

通过软件测试可以发现和消除（调试）许多错误。然而，软件测试（如果有的话）很少能够消除所有的错误；有些程序员说，"每一个程序至少都有一个错误"。在软件开发的瀑布方法中，通常使用独立的测试团队，但在较新的方法中，统称为敏捷软件开发，开发者经常亲自做所有的测试，并定期向用户/客户展示该软件以获得反馈。软件可以通过单元测试、回归测试等方法进行测试，可以手工完成，因为要测试的代码量可能相当大，所以最常见的是自动进行测试。例如，美国航空航天局（NASA）拥有许多极为严格的操作系统和通信功能的软件测试程序。许多基于 NASA 的操作通过命令程序交互，相互识别。这使很多在 NASA 工作的人能够检查和评估系统的整体功能。包含命令软件的程序使硬件工程和系统操作能够更容易地共同发挥其功能。

Unit 3

Text A

Operating System

1. What is an operating system?

An operating system is the core software component of your computer. It performs many functions and is, in very basic terms, an interface between your computer and the outside world. In the section about hardware, a computer is described as consisting of several component parts including your monitor, keyboard, mouse, and other parts. The operating system provides an interface to these parts using what is referred to as "drivers". This is why sometimes when you install a new printer or other piece of hardware, your system will ask you to install more software called a driver.

2. What does a driver do?

A driver is a specially written program which understands the operation of the device it interfaces to, such as a printer, video card, sound card or CD-ROM drive. It translates commands from the operating system or user into commands understood by the component part it interfaces with. It also translates responses from the component part back to responses that can be understood by the operating system, application program, or user.

3. Other operating system functions

The operating system provides for several other functions including:
- System tools (programs) used to monitor computer performance, debug problems, or maintain parts of the system.
- A set of libraries or functions which programs may use to perform specific tasks especially relating to interfacing with computer system components.

The operating system makes these interfacing functions (see Figure 3-1) along with its other functions operate smoothly and these functions are mostly transparent to the user.

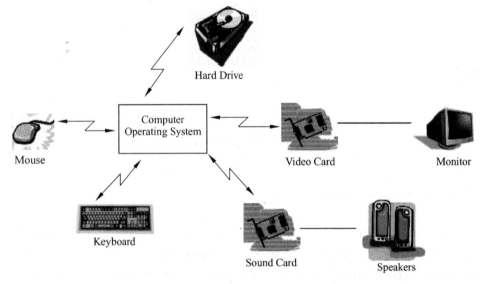

Figure 3-1　Operating System Interfaces

4. Operating system concerns

As mentioned previously, an operating system is a computer program. Operating systems are written by human programmers who can make mistakes. Therefore, there can be errors in the code even though there may be some testing before the product is released. Some companies have better software quality control and testing than others, so you may notice varying levels of quality from operating system to operating system. Errors in operating systems cause three main types of problems:
- System crashes and instabilities — These can happen due to a software bug typically in the operating system, although computer programs being run on the operating system can make the system more unstable or may even crash the system by themselves. This varies depending on the type of operating system. A system crash is

the act of a system freezing and becoming unresponsive which would cause the user to need to reboot.
- Security flaws — Some software errors leave a door open for the system to be broken into by unauthorized intruders. As these flaws are discovered, unauthorized intruders may try to use these to gain illegal access to your system. Patching these flaws often will help keep your computer system secure.
- Malfunctions — Sometimes errors in the operating system will cause the computer not to work correctly with some peripheral devices such as printers.

5. Operating system types

Let us look at the different types of operating systems and know how they differ from one another.

- Real-time operating system

It is a multitasking operating system that aims at executing real-time applications. Real-time operating systems often use specialized scheduling algorithms so that they can achieve a deterministic nature of behavior. The main object of real-time operating systems is their quick and predictable response to events. They either have an event-driven design or a time-sharing one. An event-driven system switches between tasks based on their priorities while time-sharing operating systems switch tasks based on clock interrupts.

- Multi-user and single-user operating systems

Multi-user computer operating systems allow multiple users to access a computer system simultaneously. Time-sharing systems can be classified as multi-user systems as they enable a multiple user access to a computer through time sharing. Single-user operating systems, as opposed to a multi-user operating system, are usable by only one user at a time. Being able to have multiple accounts on a Windows operating system does not make it a multi-user system. Rather, only the network administrator is the real user. But for a Unix-like operating system, it is possible for two users to log in at a time and this capability of the OS makes it a multi-user operating system.

- Multi-tasking and single-tasking operating systems

When a single program is allowed to run at a time, the system is grouped under the single-tasking system category, while in case the operating system allows for execution of multiple tasks at a time, it is classified as a multi-tasking operating system. Multi-tasking can be of two types, namely preemptive and cooperative. In preemptive multitasking, the operating system slices the CPU time and dedicates one slot to each of the programs. Unix-like operating systems such as Solaris and Linux support preemptive multitasking. If you are aware of the multithreading terminology, you can consider this type of multi-tasking as similar to interleaved multithreading. Cooperative multitasking is achieved by relying on each process to give time to the other processes in a defined manner. This kind of multitasking is similar to the idea of block multithreading in which one thread runs till it is blocked by some other event.

- Distributed operating system

An operating system that manages a group of independent computers and makes them appear to be a single computer is known as a distributed operating system. The development of networked computers that could be linked and made to communicate with each other gave rise to distributed computing. Distributed computations are carried out on more than one machine. When computers in a group work in cooperation, they make a distributed system.

- Embedded operating system

The operating systems designed for being used in embedded computer systems are known as embedded operating systems. They are designed to operate on small machines like PDAs with less autonomy. They are able to operate with a limited number of resources. They are very compact and extremely efficient by design.

- Mobile operating system

Though not a functionally distinct kind of operating system, mobile OS is definitely an important mention in the list of operating system types. A mobile OS controls a mobile device and its design supports wireless communication and mobile applications. It has built-in support for mobile multimedia formats. Tablet PCs and smartphones run on mobile operating systems.

- Batch processing and interactive systems

Batch processing refers to execution of computer programs in "batches" without manual intervention. In batch processing systems, programs are collected, grouped and processed on a later date. There is no prompting the user for inputs as input data are collected in advance for future processing. Input data are collected and processed in batches, hence the name batch processing. IBM's z/OS has batch processing capabilities. As against this, interactive operating requires user intervention. The process cannot be executed in the user's absence.

- online and offline processing systems

In online processing of data, the user remains in contact with the computer and processes are executed under control of the computer's central processing unit. When processes are not executed under direct control of the CPU, the processing is referred to as offline. Let's take the example of batch processing. Here, the batching or grouping of data can be done without user and CPU intervention; it can be done offline. But the actual process execution may happen under direct control of the processor, that is online.

New Words

core	[kɔː]	n.核心
install	[inˈstɔːl]	vt.安装，安置
translate	[trænsˈleit]	vt.翻译，解释，转化
response	[risˈpɔns]	n.回答，响应，反应
smoothly	[ˈsmuːðli]	adv.平稳地
transparent	[trænsˈpɛərənt]	adj.透明的，显然的，明晰的
mention	[ˈmenʃən]	n.& v.论及，提及
mistake	[misˈteik]	n.错误，过失
		v.弄错，误解
release	[riˈliːs]	vt. & n.发布
control	[kənˈtrəul]	n. & vt.控制，支配
instability	[instəˈbiliti]	n.不稳固，不稳定
unstable	[ʌnˈsteibl]	adj.不牢固的，不稳定的
freezing	[ˈfriːziŋ]	adj.冻结的
unresponsive	[ʌnriˈspɔnsiv]	adj.无反应的，没有回答的
reboot	[riːˈbuːt]	n.重新启动
security	[siˈkjuəriti]	n.安全
flaw	[flɔː]	n.缺陷；裂痕
unauthorized	[ʌnˈɔːθəraizd]	adj.未经授权的，未经许可的，未经批准的
intruder	[inˈtruːdə]	n.入侵者
illegal	[iˈliːgəl]	adj.不合法的，违法的
malfunction	[mælˈfʌŋkʃən]	n.故障，失灵，功能失常
algorithm	[ˈælgəriðəm]	n.算法
deterministic	[di͵təːmiˈnistik]	adj.确定性的
predictable	[priˈdiktəbl]	adj.可预言的
event-driven	[iˈvent-ˈdrivn]	n.事件驱动
interrupt	[͵intəˈrʌpt]	vt.中断
		n.中断信号
preemptive	[priːˈemptiv]	adj.优先的，抢先的
multitasking	[ˈmʌlti͵tɑːskiŋ]	n.多任务处理
slice	[slais]	n.一份，部分，片段
		v.切(片)
multithreading	[ˈmʌltiˈθrediŋ]	n.多线程
thread	[θred]	n.线程
embedded	[emˈbedid]	adj.嵌入的，植入的，内含的

autonomy	[ɔːˈtɒnəmi]	n.自治
compact	[ˈkɒmpækt]	adj.紧凑的，紧密的，简洁的
definitely	[ˈdefinitli]	adv.明确地，干脆地
batch	[bætʃ]	n.批处理
interactive	[ˌintərˈæktiv]	adj.交互式的
absence	[ˈæbsəns]	n.不在，缺席，缺乏，没有

📖 Phrases

video card	视频卡
sound card	声卡
along with…	连同……一起，随同……一起
break into	攻入，破门而入，侵占
a member of	一个成员
come out	出现
lie in	存在于
real-time operating system	实时操作系统
multitasking operating system	多任务操作系统
aim at	瞄准，针对
scheduling algorithm	调度算法
time-sharing operating system	分时操作系统
clock interrupt	时钟中断
multi-user operating system	多用户操作系统
single-user operating system	单用户操作系统
time-sharing system	分时系统
be classified as …	被分为……
single-tasking operating system	单任务操作系统
distributed operating system	分布式操作系统
embedded computer system	嵌入式计算机系统
mobile operating system	移动操作系统
wireless communication	无线通信
batch processing	批处理
online processing system	在线处理系统
offline processing system	离线处理系统

Abbreviations

CD (Compact Disc)　　　　　　　光盘
PDA (Personal Digital Assistant)　　个人数字助理
OS (Operating System)　　　　　　操作系统

Notes

[1] It translates commands from the operating system or user into commands understood by the component part it interfaces with.

本句中，from the operating system or user 是一个介词短语，作定语，修饰和限定它前面的 commands。understood by the component part it interfaces with 是一个过去分词短语，作定语，修饰和限定它前面的 commands。在该过去分词短语中，it interfaces with 是一个定语从句，修饰和限定 the component part。translates…into…的意思是"把……翻译成……"。

[2] The operating system makes these interfacing functions along with its other functions operate smoothly and these functions are mostly transparent to the user.

本句中，The operating system 作主语，makes 作谓语，these interfacing functions along with its other functions 作宾语，operate smoothly 是一个不带 to 的动词不定式短语，作宾语补足语。

英语中，当 make、let、have、see、hear、watch、notice、feel 等动词后面用不定式作宾语补足语时，不定式都不带 to。这一点特别重要。请看下例：

I often hear people talk about this kind of printer.
我经常听人们谈论这种打印机。
Please don't forget to have him help you with your computing.
请别忘了让他帮你做运算。

[3] These can happen due to a software bug typically in the operating system, although computer programs being run on the operating system can make the system more unstable or may even crash the system by themselves.

本句中，due to a software bug typically in the operating system 是一个原因状语从句。due to 的意思是"由于，因为"。although computer programs being run on the operating system can make the system more unstable or may even crash the system by themselves 是一个让步状语从句。在该从句中，computer programs 作主语，being run on the operating system 作定语，修饰 computer programs，can make 作谓语，the system 作宾语，more unstable 作宾语补足语，or 是连词，连接并列谓语，意思是"或者"。

[4] When a single program is allowed to run at a time, the system is grouped under the single-tasking system category, while in case the operating system allows for execution of multiple tasks at a time, it is classified as a multi-tasking operating system.

本句中，while 表示对比，意思是"而，但是"，它连接了两个复合句。这两个复合句分别解释了什么是 single-tasking system，什么是 multi-tasking operating system。

英语中，while 在不同的语境中所表达的意思不同。请看下例：

While the discussion was still going on, George came in.

当讨论还在进行时，乔治走了进来。（当……时；表示时间）

Multi-user computer operating systems allow multiple users to access a computer system simultaneously while single-user operating systems are usable by only one user at a time.

多用户计算机操作系统允许多个用户同时访问一个计算机系统，而单用户操作系统一次只能被一个用户使用。（而，但是；表示对比）

While this printer is of good quality, I think it is too expensive.

尽管这台打印机质量很好，但我认为还是太贵了。（虽然，尽管；表示让步）

We can surely overcome these difficulties while we make our best. （只要；表示条件）

只要我们竭尽全力，就一定能克服这些困难。

[5] An operating system that manages a group of independent computers and makes them appear to be a single computer is known as a distributed operating system.

本句中，that manages a group of independent computers and makes them appear to be a single computer 是一个定语从句，修饰和限定 An operating system。be known as 的意思是"被称为……"。

Exercises

【Ex. 1】 根据课文内容回答问题。

1. What is an operating system?
2. What does a driver do?
3. What are system tools (programs) used to?
4. Why can there be errors in the code even though there may be some testing before the product is released?
5. What are the three main types of problems errors in operating systems cause?
6. What is a system crash?
7. What may happen if there are security flaws? What should we do?
8. Are there many types of operating systems? What is the most common one?
9. What can time-sharing systems be classified as? What is the difference between them?
10. What are embedded operating systems? What are they designed to do?

【Ex. 2】 根据下面的英文解释，写出相应的英文词汇。
1. _____: To set in place and prepare for operation.
2. _____: A signal that initiates an operation defined by an instruction.
3. _____: In programming, to convert a program from one language to another.
4. _____: An error or a fault resulting from defective judgment, deficient knowledge, or carelessness.
5. _____: A particular version of a piece of software, most commonly associated with the most recent version.
6. _____: Management of a computer and its processing abilities so as to maintain order as tasks and activities are carried out.
7. _____: For a system or program, to fail to function correctly, resulting in the suspension of operation.
8. _____: The quality or condition of being erratic or undependable.
9. _____: To turn a computer off and then on again; restart the operating system.
10. _____: A combination of input, output, and computing hardware that can be used for work by an individual.

【Ex. 3】 把下列句子翻译为中文。
1. It loads the operating system into memory and allows it to begin operation.
2. On the computer, there are two basic types of items that need to be organized.
3. Fonts are used by computer for on-screen display and printers for hardcopy output.
4. Optical fiber is thin filaments of glass through which light beams are transmitted.
5. When you type things on the keyboard, the letters and numbers show up on the monitor.
6. An intranet is a private network. There are many intranets scattered all over the world.
7. On the computer screen, a folder most often looks like a yellow or blue paper file folder.
8. Once you've encoded your source content, the process of creating streaming media is complete.
9. Syntactically, a domain name consists of a sequence of names (labels) separated by periods (dots).
10. The quality of video you see on your monitor depends on both the video card and the monitor you choose.

【Ex. 4】 将下列词填入适当的位置（每词只用一次）。

| graphical | through | applications | defined | interact |
| boot | command | use | loaded | requests |

An operating system (sometimes abbreviated as "OS") is the program that, after being initially __(1)__ into the computer by a __(2)__ program, manages all the other programs in a computer. The other programs are called __(3)__ or application programs. The application programs make __(4)__ of the operating system by making __(5)__ for services through a __(6)__ application program interface (API). In addition, users can __(7)__ directly with the operating system __(8)__ a user interface such as a __(9)__ language or a __(10)__ user interface (GUI).

Text B

ETL

In computing, Extract, Transform, Load (ETL) refers to a process in database usage and especially in data warehousing. The ETL process became a popular concept in the 1970s. Data extraction is where data is extracted from homogeneous or heterogeneous data sources; data transformation where the data is transformed for storing in the proper format or structure for the purposes of querying and analysis; data loading where the data is loaded into the final target database, more specifically, an operational data store, data mart, or data warehouse.

Since data extraction takes time, it is common to execute the three phases in parallel. While the data is being extracted, another transformation process executes while processing the data already received and prepares it for loading while the data loading begins without waiting for the completion of the previous phases.

ETL systems commonly integrate data from multiple applications (systems), typically developed and supported by different vendors or hosted on separate computer hardware. The disparate systems containing the original data are frequently managed and operated by different employees. For example, a cost accounting system may combine data from payroll, sales, and purchasing.

1. Extract

The first part of an ETL process involves extracting the data from the source system(s). In many cases, this represents the most important aspect of ETL, since extracting data correctly sets the stage for the success of subsequent processes. Most data-warehousing projects combine data from different source systems. Each separate system may also use a different data organization and/or format. Common data-source formats include relational databases, XML and flat files, but they may also include non-relational database structures such as Information Management System (IMS) or other data structures such as Virtual Storage Access Method (VSAM) or Indexed Sequential Access Method (ISAM), or even formats fetched from outside sources by means such as web spidering or screen scraping. The streaming of the extracted data source and loading on-the-fly to the destination database is another way of performing ETL when no intermediate data storage is required. In general, the extraction phase aims to convert the data into a single format appropriate for transformation processing.

An intrinsic part of the extraction involves data validation to confirm whether the data pulled from the sources has the correct/expected values in a given domain (such as a pattern/default or list of values). If the data fails the validation rules it is rejected entirely or in part. The rejected data is ideally reported back to the source system for further analysis to identify and to rectify the incorrect records. In some cases, the extraction process itself may have to do a data-validation rule in order to accept the data and flow to the next phase.

2. Transform

In the data transformation stage, a series of rules or functions are applied to the extracted data in order to prepare it for loading into the end target. Some data does not require any transformation at all; such data is known as "direct move" or "pass through" data.

An important function of transformation is the cleaning of data, which aims to pass only "proper" data to the target. The challenge when different systems interact is in the relevant systems' interfacing and communicating. Character sets that may be available in one system may not be so in others.

In other cases, one or more of the following transformation types may be required to meet the business and technical needs of the server or data warehouse:

- Selecting only certain columns to load: (or selecting null columns not to load). For example, if the source data has three columns (aka "attributes"), roll_no, age, and salary, then the selection may take only roll_no and salary. Or, the selection mechanism may ignore all those records where salary is not present (salary = null).
- Translating coded values: (*e.g.*, if the source system codes male as "1" and female as "2", but the warehouse codes male as "M" and female as "F")
- Encoding free-form values: (*e.g.*, mapping "Male" to "M")
- Deriving a new calculated value: (*e.g.*, sale_amount = qty * unit_price)
- Sorting or ordering the data based on a list of columns to improve search performance
- Joining data from multiple sources (*e.g.*, lookup, merge) and deduplicating the data
- Aggregating (for example, rollup—summarizing multiple rows of data—total sales for each store, and for each region, etc.)

- Generating surrogate-key values
- Transposing or pivoting (turning multiple columns into multiple rows or vice versa)
- Splitting a column into multiple columns (*e.g.*, converting a comma-separated list, specified as a string in one column, into individual values in different columns)
- Disaggregating repeating columns
- Looking up and validating the relevant data from tables or referential files
- Applying any form of data validation; failed validation may result in a full rejection of the data, partial rejection, or no rejection at all, and thus none, some, or all of the data is handed over to the next step depending on the rule design and exception handling; many of the above transformations may result in exceptions, e.g., when a code translation parses an unknown code in the extracted data

3. Load

The load phase loads the data into the end target, which may be a simple delimited flat file or a data warehouse. Depending on the requirements of the organization, this process varies widely. Some data warehouses may overwrite existing information with cumulative information; updating extracted data is frequently done on a daily, weekly, or monthly basis. Other data warehouses (or even other parts of the same data warehouse) may add new data in a historical form at regular intervals—for example, hourly. To understand this, consider a data warehouse that is required to maintain sales records of the last year. This data warehouse overwrites any data older than a year with newer data. However, the entry of data for any one year window is made in a historical manner. The timing and scope to replace or append are strategic design choices dependent on the time available and the business needs. More complex systems can maintain a history and audit trail of all changes to the data loaded in the data warehouse.

As the load phase interacts with a database, the constraints defined in the database schema—as well as in triggers activated upon data load—apply (for example, uniqueness, referential integrity, mandatory fields), which also contribute to the overall data quality performance of the ETL process.

- For example, a financial institution might have information on a customer in several departments and each department might have that customer's information listed in a different way. The membership department might list the customer by name, whereas the accounting department might list the customer by number. ETL can bundle all of these data elements and consolidate them into a uniform presentation, such as for storing in a database or data warehouse.
- Another way that companies use ETL is to move information to another application

permanently. For instance, the new application might use another database vendor and most likely a very different database schema. ETL can be used to transform the data into a format suitable for the new application to use.

- An example would be an Expense and Cost Recovery System (ECRS) such as used by accountancies, consultancies, and legal firms. The data usually ends up in the time and billing system, although some businesses may also utilize the raw data for employee productivity reports to Human Resources (personnel dept.) or equipment usage reports to Facilities Management.

4. Real-life ETL cycle

The typical real-life ETL cycle consists of the following execution steps:

(1) Cycle initiation

(2) Build reference data

(3) Extract (from sources)

(4) Validate

(5) Transform (clean, apply business rules, check for data integrity, create aggregates or disaggregates)

(6) Stage (load into staging tables, if used)

(7) Audit reports (for example, on compliance with business rules. Also, in case of failure, helps to diagnose/repair)

(8) Publish (to target tables)

(9) Archive

5. Challenges

ETL processes can involve considerable complexity, and significant operational problems can occur with improperly designed ETL systems.

The range of data values or data quality in an operational system may exceed the expectations of designers at the time validation and transformation rules are specified. Data profiling of a source during data analysis can identify the data conditions that must be managed by transform rules specifications, leading to an amendment of validation rules explicitly and implicitly implemented in the ETL process.

Data warehouses are typically assembled from a variety of data sources with different formats and purposes. As such, ETL is a key process to bring all the data together in a standard, homogeneous environment.

Design analysis should establish the scalability of an ETL system across the lifetime of its usage — including understanding the volumes of data that must be processed within service level agreements. The time available to extract from source systems may change, which may mean the same amount of data may have to be processed in less time. Some ETL systems have to scale to process terabytes of data to update data warehouses with tens of terabytes of data. Increasing volumes of data may require designs that can scale from daily batch to multiple-day micro batch to integration with message queues or real-time change data capture for continuous transformation and update.

6. Performance

ETL vendors benchmark their record-systems at multiple TB (terabytes) per hour (or ~ 1 GB per second) using powerful servers with multiple CPUs, multiple hard drives, multiple gigabit-network connections, and lots of memory.

In real life, the slowest part of an ETL process usually occurs in the database load phase. Databases may perform slowly because they have to take care of concurrency, integrity maintenance, and indices. Thus, for better performance, it may make sense to employ:
- Direct Path Extract method or bulk unload whenever is possible (instead of querying the database) to reduce the load on source system while getting high speed extract;
- Most of the transformation processing outside of the database;
- Bulk load operations whenever possible.

Still, even using bulk operations, database access is usually the bottleneck in the ETL process. Some common methods used to increase performance are:
- Partition tables (and indices): try to keep partitions similar in size (watch for null values that can skew the partitioning);
- Do all validation in the ETL layer before the load: disable integrity checking (disable constraint...) in the target database tables during the load;
- Disable triggers (disable trigger...) in the target database tables during the load: simulate their effect as a separate step;
- Generate IDs in the ETL layer (not in the database);
- Drop the indices (on a table or partition) before the load and recreate them after the load (SQL: drop index...; create index...);
- Use parallel bulk load when possible—works well when the table is partitioned or there are no indices (Note: attempt to do parallel loads into the same table (partition) usually causes locks—if not on the data rows, then on indices);
- If a requirement exists to do insertions, updates, or deletions, find out which rows should be processed in which way in the ETL layer, and then process these three operations in the database separately; you often can do bulk load for inserts, but

updates and deletes commonly go through an API (using SQL).

Whether to do certain operations in the database or outside may involve a trade off. For example, removing duplicates using distinct may be slow in the database; thus, it makes sense to do it outside. On the other side, if using distinct significantly (x100) decreases the number of rows to be extracted, then it makes sense to remove duplications as early as possible in the database before unloading data.

A common source of problems in ETL is a big number of dependencies among ETL jobs. For example, job "B" cannot start while job "A" is not finished. One can usually achieve better performance by visualizing all processes on a graph, and trying to reduce the graph making maximum use of parallelism, and making "chains" of consecutive processing as short as possible. Again, partitioning of big tables and their indices can really help.

Another common issue occurs when the data are spread among several databases, and processing is done in those databases sequentially. Sometimes database replication may be involved as a method of copying data between databases, but it can significantly slow down the whole process. The common solution is to reduce the processing graph to only three layers:

- Sources
- Central ETL layer
- Targets

This approach allows processing to take maximum advantage of parallelism. For example, if you need to load data into two databases, you can run the loads in parallel (instead of loading into the first, and then replicating into the second).

Sometimes processing must take place sequentially. For example, dimensional (reference) data are needed before one can get and validate the rows for main "fact" tables.

✎ New Words

load	[ləud]	vt.装载，加载，装填
		n.负荷，装载量，工作量，负载，加载
homogeneous	[ˌhɔməu'dʒi:niəs]	adj.同类的，相似的
heterogeneous	[ˌhetərəu'dʒi:niəs]	adj.不同种类的，异类的
store	['stɔ:]	vt.存储，保管
format	['fɔ:mæt]	n.格式，形式
parallel	['pærəlel]	adj.平行的，并联的
		v.并行，平行
integrate	['intigreit]	vt.集成，使成整体，使一体化
represent	[ˌrepri'zent]	vt.表现，扮演

subsequent	[ˈsʌbsikwənt]	adj.后来的，并发的
combine	[kəmˈbain]	v.（使）联合，（使）结合
streaming	[ˈstriːmiŋ]	n.流
intermediate	[ˌintəˈmiːdiət]	adj.中间的
intrinsic	[inˈtrinsik]	adj.（指价值、性质）固有的，内在的，本质的
validation	[væliˈdeiʃən]	n.确认，有效
default	[diˈfɔːlt]	n.默认（值），缺省（值）
reject	[riˈdʒekt]	vt.拒绝，抵制，丢弃
rectify	[ˈrektifai]	vt.矫正，调整
rule	[ruːl]	n.规则，惯例，准则，标准
		vt.规定，统治，支配
interact	[ˌintərˈækt]	vi.互相作用，互相影响
column	[ˈkɔləm]	n.栏，列
mechanism	[ˈmekənizəm]	n.机制
ignore	[igˈnɔː]	vt.不理睬，忽视
encode	[inˈkəud]	vt.编码
sort	[sɔːt]	vt.排序
order	[ˈɔːdə]	vt.排序，分类
join	[dʒɔin]	vt.连接
lookup	[ˈlukʌp]	v.查找
duplicate	[ˈdjuːplikeit]	vt.复制，重复
pivot	[ˈpivət]	vi.转置
disaggregate	[disˈægrigeit]	v.去除，分解
rejection	[riˈdʒekʃən]	n.拒绝，丢弃
handling	[ˈhændliŋ]	n.处理
		adj.操作的
delimit	[diˈlimit]	vt.定界限，划界
overwrite	[ˌəuvəˈrait]	v.改写，覆盖
cumulative	[ˈkjuːmjulətiv]	adj.累积的
trigger	[ˈtrigə]	vt.引发，引起，触发
uniqueness	[juˈniːknis]	n.唯一性，单值性，独特性
integrity	[inˈtegriti]	n.完整性
permanently	[ˈpəːməntli]	adv.永存地，不变地
cycle	[ˈsaikl]	n.周期，循环
		vi.循环
validate	[ˈvælideit]	vt.确认，证实，验证
compliance	[kəmˈplaiəns]	n.合规，依从

diagnose	['daiəgnəuz]	v.诊断
repair	[ri'pɛə]	n.修理，修补 vt.修理，修补，补救，纠正
considerable	[kən'sidərəbl]	adj.相当大（或多）的，相当可观的
improperly	[im'prɔpəli]	adv.不正确地，不适当地
designer	[di'zainə]	n.设计者
specify	['spesifai]	vt.指定
condition	[kən'diʃən]	n.条件，情形，环境 vt.以……为条件，使达到要求的情况
specification	[ˌspesifi'keiʃən]	n.规范，规格，说明书
amendment	[ə'mendmənt]	n.改善，改正
scalability	[ˌskeilə'biliti]	n.可量测性
agreement	[ə'gri:mənt]	n.协定，协议
continuous	[kən'tinjuəs]	adj.连续的，持续的
bulk	[bʌlk]	n.大批，大多数；散装 vt.显得大，显得重要
unload	[ˌʌn'ləud]	v.卸载
bottleneck	['bɔtlnek]	n.瓶颈
disable	[dis'eibl]	v.使无效，使失去能力
insertion	[in'sə:ʃən]	n.插入
parallelism	['pærəlelizəm]	n.平行，并行
replication	[ˌrepli'keiʃən]	n.复制
dimensional	[di'menʃənəl]	adj.维的，空间的

Phrases

refer to	指的是；涉及；适用于
data warehousing	数据入库，数据存入
data extraction	数据提取
data source	数据源
data transformation	数据变换，数据转换
data loading	数据载入，数据装载
operational data store	操作型数据存储
data mart	数据集市，数据市场
take time	费时
original data	原始数据
data organization	数据结构，数据组织

relational database	关系数据库
flat file	平面文件
non-relational database	非关系型数据库
web spidering	网页蜘蛛，网页爬虫，网页搜索
screen scraping	屏幕抓取
on-the-fly	在不停机状态下，即时
convert … into …	把……转换为……
appropriate for	适于，合乎
data validation	数据有效性
a series of	一连串的，一系列的
be applied to	适用于，应用于，施加于
character set	字符集
calculated value	计算值
look up	查找
hand over	移交
audit trail	审计跟踪
mandatory field	必备字段，必填字段
data element	数据元素
be suitable for …	适合……的
reference data	参考数据
data integrity	数据完整性
staging table	临时表
transformation rule	转换规则
data profiling	数据剖析
message queue	消息队列
change data capture	变更数据捕获
partition table	分区表
trade off	权衡
consecutive processing	串行处理，顺序处理，连续处理

❧ Abbreviations

ETL (Extract, Transform, Load)	抽取、转换、加载
XML (eXtensible Markup Language)	可扩展标记语言
VSAM (Virtual Storage Access Method)	虚拟存储存取方法，虚拟存储访问方法
ISAM (Indexed Sequential Access Method)	索引顺序存取方法，索引顺序访问方法
ECRS (Expense and Cost Recovery System)	费用与成本回收系统

API (Application Programming Interface)　　　应用程序编程接口

Exercises

【Ex. 5】 根据课文内容回答问题。
1. What does ETL stand for? What does it refer to in computing?
2. What does ETL systems commonly do?
3. What does the first part of an ETL process involve?
4. Why does extracting data represent the most important aspect of ETL in many cases?
5. What does the extraction phase aim to do in general?
6. What happens if the data fails the validation rules?
7. What is an important function of transformation?
8. What does the load phase do?
9. Where does the slowest part of an ETL process usually occur in real life? Why?
10. What is a common source of problems in ETL?

参考译文

操 作 系 统

1. 什么是操作系统

操作系统是计算机的核心软件。它执行许多功能，用很基本的术语说，它是计算机与外部设备之间的接口。在硬件一节中，计算机被描述为由许多独立的部件组成，包括显示器、键盘、鼠标及其他部件。操作系统使用所谓的"驱动程序"给这些部件提供接口。这就是为什么当你安装一个新打印机或其他硬件时，系统会问你是否进一步安装称作驱动程序的软件。

2. 驱动程序做什么

驱动程序是一个经过编写的特殊程序，它了解与其接口的设备（如打印机、显卡、声卡或光盘驱动器）的操作。它把来自操作系统或用户的命令翻译为其接口的设备可以理解的命令。它也把来自这些部件的响应翻译为操作系统、应用软件或用户可以理解的响应。

3. 其他操作系统的功能

操作系统的其他功能包括：
- 用于监控计算机执行、排除故障或维护系统部件的系统工具。
- 程序用来执行特殊任务、特别是与计算机系统部件接口相关任务的一系列功能库或函数。

操作系统使这些接口功能及其他功能平稳运行，并且这些功能对用户通常是透明的。

4. 操作系统关注的事项

如上所述，操作系统是计算机程序。操作系统由可能出错的程序员编写。因此，即使在发布前进行了测试，它仍可能有一些错误代码。有些公司的软件质量控制和测试优于其他公司，所以也许是注意到了不同的操作系统质量不同。操作系统的错误引起以下3类主要问题：
- 系统崩溃和不稳定性——这通常由操作系统中的软件错误引起，尽管运行在操作系统上的计算机程序可以使系统更不稳定，甚至由它们引起系统崩溃。这些变化取决于操作系统的类型。系统崩溃是系统冻结并且没有反应的行为，用户必须重新启动。
- 安全漏洞——某些软件错误为未经授权的入侵者打开进入系统的大门。由于这些漏洞，未经授权的入侵者也许试图使用它们非法访问你的系统。给这些漏洞打补丁通常可以使计算机系统变得安全。
- 功能失常——有时操作系统的错误可以引起计算机及一些外设（如打印机）不能正常工作。

5. 操作系统的类型

让我们来看看不同类型的操作系统并了解它们之间的区别。
- 实时操作系统

它是一个多任务操作系统，其目的是执行实时应用。实时操作系统通常使用专门的调度算法，以便可以实现确定性的行为。实时操作系统的主要目的是对事件做出快速和可预测的响应。其设计或者是事件驱动的或者是分时的。事件驱动的系统基于优先级在任务之间切换，而分时操作系统基于时钟中断在任务之间切换。

- 多用户和单用户的操作系统

多用户计算机操作系统允许多个用户同时访问计算机系统。分时系统可以划分为多用户系统和单用户操作系统。多用户操作系统通过分时让多个用户访问一个计算机。与其相对应的单用户操作系统一次只有一个用户使用。Windows 操作系统可能有多个账户但并不是一个多用户系统。只有网络管理员是真正的用户。但对于一个类 UNIX 操作系统，它可以让两个用户在同一时间登录，这种性能使其成为多用户操作系统。

- 多任务处理及单任务操作系统

当一次只能运行一个程序时，该系统仍归于单任务系统。在操作系统允许同时执行多任务的情况下，它就归于多任务操作系统。多任务处理可以分为两类：抢先型与协作型。在抢先多任务操作系统中，系统为每个程序切分一段 CPU 时隙。类 UNIX 的操作系统（如 Solaris 和 Linux）支持抢先式多任务。如果你了解多线程，那么可以把这种类型的多任务处理能力当作交错的多线程。协作多任务通过每一个进程以确定的方式给其他进程分配时间来实现。这种多任务类似于块多线程的想法，在一个线程中运行，直到它被另外一些事件闭锁。

- 分布式操作系统

管理一组独立的计算机并使得它们看起来是一台计算机，这就被称为分布式操作系统。可以链接并彼此通信的计算机的发展带来了分布式计算。分布式计算在多个计算机上进行。当一组计算机协作工作时就构成一个分布式系统。

- 嵌入式操作系统

嵌入式操作系统设计用于嵌入式计算机系统。它们为像 PDA 这类更少自治的小型机而设计。它们能够在资源有限的系统中运行。它们非常紧凑，设计效率非常高。

- 移动操作系统

移动操作系统虽然在功能上与其他操作系统并没有明显的不同，但绝对是操作系统类型列表中的重要一项。移动操作系统控制移动设备，其设计支持无线通信和移动应用。它内置支持移动多媒体格式。平板电脑和智能手机都运行在移动操作系统上。

- 批处理和交互系统

批处理指的是"按批"执行计算机程序，无需人工干预。在批处理系统中，收集程序、分组并在稍后的日期处理。并不提示用户输入数据，因为以后要处理的数据已经提前收集了。因为输入数据分批收集和处理故名批处理。 IBM 的 z/OS 具有批处理能力。与此相对，交互式的操作需要用户干预。用户不在就不能执行。

- 在线和离线处理系统

在线数据处理时，用户保持与计算机的联系并在计算机中央处理单元的控制下执行。当进程不在 CPU 的直接控制下执行时，该处理被称为离线。让我们以批处理为例介绍。这里，数据的分批或分组可以无须用户和 CPU 的干预，它可以离线完成。但实际执行过程中可以在处理器直接控制下完成，也就是在线完成。

Unit 4

Text A

R Programming Language

R is a programming language and software environment for statistical computing and graphics supported by the R Foundation for Statistical Computing. The R language is widely used among statisticians and data miners for developing statistical software and data analysis. Polls, surveys of data miners, and studies of scholarly literature databases show that R's popularity has increased substantially in recent years.

R is a GNU project. The source code for the R software environment is written primarily in C, Fortran, and R. R is freely available under the GNU General Public License, and precompiled binary versions are provided for various operating systems. While R has a command line interface, there are several graphical front-ends available.

1. History

R is an implementation of the S programming language combined with lexical scoping semantics inspired by Scheme. S was created by John Chambers while at Bell Labs. There are some important differences, but much of the code written for S runs unaltered.

R was created by Ross Ihaka and Robert Gentleman at the University of Auckland, New Zealand, and is currently developed by the R Development Core Team, of which Chambers is a member. The project was conceived in 1992, with an initial version released in 1994 and a stable beta version in 2000.

2. Statistical features

R and its libraries implement a wide variety of statistical and graphical techniques, including linear and nonlinear modeling, classical statistical tests, time-series analysis, classification, clustering, and others. R is easily extensible through functions and extensions, and the R community is noted for its active contributions in terms of packages. Many of R's standard functions are written in R itself, which makes it easy for users to follow the algorithmic choices made. For computationally intensive tasks, C, C++, and Fortran code can be linked and called at run time. Advanced users can write C, C++, Java, .NET or Python code to manipulate R objects directly. R is highly extensible through the use of user-submitted packages for specific functions or specific areas of study. Due to its S heritage, R has stronger object-oriented programming facilities than most statistical computing languages. Extending R is also eased by its lexical scoping rules.

Another strength of R is static graphics, which can produce publication-quality graphs, including mathematical symbols. Dynamic and interactive graphics are available through additional packages.

R has Rd, its own LaTeX-like documentation format, which is used to supply comprehensive documentation, both on-line in a number of formats and in hard copy.

3. Programming features

R is an interpreted language, and users typically access it through a command-line interpreter. If a user types 2+2 at the R command prompt and presses enter, the computer replies with 4.

Like other similar languages such as APL and MATLAB, R supports matrix arithmetic. R's data structures include vectors, matrices, arrays, data frames (similar to tables in a relational database) and lists. R's extensible object system includes objects for (among others): regression models, time-series and geo-spatial coordinates. The scalar data type was never a data structure of R. Instead, a scalar is represented as a vector with length one.

R supports procedural programming with functions and, for some functions, object-oriented programming with generic functions. A generic function acts differently depending on the classes of arguments passed to it. In other words, the generic function dispatches the function (method) specific to that class of object. For example, R has a generic print function that can print almost every class of object in R with a simple print(objectname) syntax.

Although used mainly by statisticians and other practitioners requiring an environment for statistical computation and software development, R can also operate as a general matrix calculation toolbox—with performance benchmarks comparable to GNU Octave or MATLAB.

4. Packages

The capabilities of R are extended through user-created packages, which include specialized statistical techniques, graphical devices (such as the ggplot2 package developed by Hadley Wickham), import/export capabilities, reporting tools (knitr, Sweave), etc. These packages are developed primarily in R, and sometimes in Java, C, C++, and Fortran.

A core set of packages is included with the installation of R, with more than 7,801 additional packages (as of January 2016) available at the Comprehensive R Archive Network (CRAN), Bioconductor, Omegahat, GitHub, and other repositories.

The "Task Views" page (subject list) on the CRAN website lists a wide range of tasks (in fields such as Finance, Genetics, High Performance Computing, Machine Learning, Medical Imaging, Social Sciences and Spatial Statistics) to which R has been applied and for which packages are available. R has also been identified by the FDA as suitable for interpreting data from clinical research.

Other R package resources include Crantastic, a community site for rating and reviewing all CRAN packages, and R-Forge, a central platform for the collaborative development of R packages, R-related software, and projects. R-Forge also hosts many unpublished beta packages, and development versions of CRAN packages.

5. Interfaces

5.1 Graphical user interfaces

- Architect—cross-platform open source IDE for data science based on Eclipse and StatET.
- DataJoy—online R Editor focused on beginners to data science and collaboration.
- Deducer—GUI for menu-driven data analysis (similar to SPSS/JMP/Minitab).
- Java GUI for R—cross-platform stand-alone R terminal and editor based on Java (also known as JGR).
- Number Analytics—GUI for R based business analytics (similar to SPSS) working on the cloud.
- Rattle GUI—cross-platform GUI based on RGtk2 and specifically designed for data mining.
- R Commander—cross-platform menu-driven GUI based on tcltk (several plug-ins to

Rcmdr are also available).
- Revolution R Productivity Environment (RPE)—Revolution Analytics-provided Visual Studio-based IDE, and has plans for web based point and click interface.
- RGUI—comes with the precompiled version of R for Microsoft Windows.
- RKWard—extensible GUI and IDE for R.
- RStudio—cross-platform open source IDE (which can also be run on a remote Linux server).

5.2 Editors and IDEs

Text editors and integrated development environments (IDEs) with some support for R include: ConTEXT, Eclipse (StatET), Emacs (Emacs Speaks Statistics), LyX (modules for knitr and Sweave), Vim, jEdit, Kate, RStudio, Sublime Text, TextMate, Atom, WinEdt (R Package RWinEdt), Tinn-R, Notepad++, and Architect.

5.3 Scripting languages

R functionality has been made accessible from several scripting languages such as Python, Perl, Ruby, F# and Julia. Scripting in R itself is possible via a front-end called littler.

6. Comparison with SAS, SPSS, and Stata

The general consensus is that R compares well with other popular statistical packages, such as SAS, SPSS, and Stata. In a comparison of all basic features for a statistical software R is heads up with the best of statistical software.

In January 2015, the *New York Times* ran an article about R gaining acceptance among data analysts and presenting a threat for the market share occupied by commercial statistical packages, such as SAS.

✎ New Words

programming	['prəʊgræmɪŋ]	n.编程，程序设计
environment	[ɪn'vaɪərənmənt]	n.环境
graphics	['græfɪks]	n.图形
support	[sə'pɔːt]	vt. & n.支持，支撑
foundation	[faʊn'deɪʃən]	n.基金，基金会
analysis	[ə'nælɪsɪs]	n.分析；分解
poll	[pəʊl]	n.民意调查；投票选举；投票数
		vi.投票；做民意调查

survey	['sə:vei] [sə'vei]	n.调查，民意调查；民意测验；测量 vt.调查（收入，民意等）
scholarly	['skɔləli]	adj.学者气质的，学者风度的
literature	['litəritʃə]	n.文学（作品），文艺，著作，文献
popularity	[,pɔpju'læriti]	n.普及，流行，声望
substantially	[səb'stænʃəli]	adv.充分地
project	['prɔdʒekt]	n.项目，方案，工程
primarily	['praimərəli]	adv.首要地，主要地；首先
available	[ə'veiləbl]	adj.可用到的，可利用的，有用的
precompiled	[pri:kəm'paild]	adj.预编译的
interface	['intə(:),feis]	n.接口，界面
front-end	[frʌnt-end]	n.前端
implementation	[,implimen'teiʃən]	n.执行，实现
lexical	['leksikəl]	adj.词汇的；具词典性质的，词典的
semantics	[si'mæntiks]	n.语义；语义学
inspire	[in'spaiə]	vt.激励；鼓舞；启迪；赋予灵感
unaltered	[ʌn'ɔ:ltəd]	adj.未被改变的，不变的，照旧的
conceive	[kən'si:v]	vt.构思 vi.考虑，设想
stable	['steibl]	adj.稳定的
library	['laibrəri]	n.库
implement	['implimənt]	v.执行，实现，贯彻 n.工具，器具
linear	['liniə]	adj.线性的
nonlinear	['nɔn'liniə]	adj.非线性的
modeling	['mɔdliŋ]	n.建模
classical	['klæsikəl]	adj.古典的
classification	[,klæsifi'keiʃən]	n.分类，分级
clustering	['klʌstəriŋ]	n.聚类
function	['fʌŋkʃən]	n.功能，作用，函数
community	[kə'mju:niti]	n.社区，团体
contribution	[,kɔntri'bju:ʃən]	n.贡献
package	['pækidʒ]	n.软件包
standard	['stændəd]	adj.标准的，普遍的 n.标准
algorithmic	[,ælgə'riðəmik]	adj.算法的
computational	[,kɔmpju'teiʃənəl]	adj.计算的

单词	音标	释义
link	[liŋk]	n.链接 vt.连接，联合 vi.连接起来
manipulate	[məˈnipjuleit]	vt.操作，使用，利用
object	[ˈɔbdʒikt]	n.目标，对象
submit	[səbˈmit]	vt.提交，递交
heritage	[ˈheritidʒ]	n.遗产，继承权，传统
facility	[fəˈsiliti]	n.设备，工具
strength	[streŋθ]	n.实力，力量
symbol	[ˈsimbəl]	n.符号，记号
document	[ˈdɔkjumənt]	n.文件，文档
comprehensive	[ˌkɔmpriˈhensiv]	adj.全面的，广泛的
interpret	[inˈtə:prit]	v.解释，说明
access	[ˈækses]	n. & vt.存取，访问
prompt	[prɔmpt]	n.提示符
enter	[ˈentə]	n.输入；回车
arithmetic	[əˈriθmətik]	n.算术，算法
vector	[ˈvektə]	n.向量，矢量
matrix	[ˈmeitriks]	n.矩阵
list	[list]	n.列表，数据清单
regression	[riˈgreʃən]	n.回归
coordinate	[kəuˈɔ:dinit]	n.坐标
scalar	[ˈskeilə]	adj.数量的，标量的 n.数量，标量
dispatch	[disˈpætʃ]	vt.调度，分派，派遣
practitioner	[prækˈtiʃənə]	n.从业者
development	[diˈveləpmənt]	n.开发，发展
operate	[ˈɔpəreit]	v.操作，运转
toolbox	[ˈtu:lbɔks]	n.工具箱
benchmark	[ˈbentʃmɑ:k]	n.基准
device	[diˈvais]	n.装置，设备
import	[imˈpɔ:t]	n. & vt.输入
export	[ˈekspɔ:t]	n. & vt.输出
installation	[ˌinstəˈleiʃən]	n.安装，装置
website	[ˈwebsait]	n.网站
task	[tɑ:sk]	n.任务，作业 v.分派任务

identify	[aiˈdentifai]	vt.识别，鉴别，确定
collaborative	[kəˈlæbəreitiv]	adj.合作的，协作的
unpublished	[ˈʌnˈpʌbliʃt]	adj.未出版发行的
collaboration	[kə,læbəˈreiʃən]	n.协作
remote	[riˈməut]	adj.远程的
server	[ˈsəːvə]	n.服务器
comparison	[kəmˈpærisn]	n.比较，对照
consensus	[kənˈsensəs]	n.一致同意，多数人的意见，舆论
acceptance	[əkˈseptəns]	n.接受，承诺，容忍，赞同，相信

Phrases

programming language	程序设计语言
data miner	数据挖掘者
in recent years	最近几年中
General Public License	通用公共许可证
provide for…	为……准备
command line	命令行
combine with …	与……结合
lexical scoping semantics	词汇作用域语义
Bell Labs	贝尔实验室
beta version	测试版
classical statistical test	经典统计测试
time-series analysis	时间序列分析
in terms of	根据，按照；用……的话，在……方面
run time	运行期间，运行时间
object-oriented programming	面向对象编程
lexical scoping rule	词法作用域规则
hard copy	硬拷贝
interpreted language	解释语言
command prompt	命令提示符
data frame	数据帧
regression model	回归模型
scalar data type	标量数据类型
be represented as …	可表示为……
generic function	类函数，广函数
a set of	一组，一套

suitable for …	适合……的
graphical user interface	图形用户界面
focus on	集中
menu-driven data analysis	菜单驱动的数据分析
special issue	特刊，专号
scripting language	脚本语言
market share	市场份额，市场占有率

Abbreviations

GNU	是"GNU is Not Unix"的递归缩写
CRAN (Comprehensive R Archive Network)	R 综合归档网
FDA (Food and Drug Administration)	（美国）食品及药物管理局
IDE (Integrated Development Environment)	集成开发环境

Notes

[1] Many of R's standard functions are written in R itself, which makes it easy for users to follow the algorithmic choices made.

本句中，which makes it easy for users to follow the algorithmic choices made 是一个非限定性定语从句，对 Many of R's standard functions are written in R itself 进行补充说明。在该非限定性定语从句中，it 是形式宾语，真正的宾语是动词不定式短语 to follow the algorithmic choices made。

[2] Another strength of R is static graphics, which can produce publication-quality graphs, including mathematical symbols.

本句中，which can produce publication-quality graphs, including mathematical symbols 是一个非限定性定语从句，对 static graphics 进行补充说明。

[3] Although used mainly by statisticians and other practitioners requiring an environment for statistical computation and software development, R can also operate as a general matrix calculation toolbox — with performance benchmarks comparable to GNU Octave or MATLAB.

本句中，Although used mainly by statisticians and other practitioners requiring an environment for statistical computation and software development 是一个过去分词短语，作状语。

[4] The capabilities of R are extended through user-created packages, which include specialized statistical techniques, graphical devices (such as the ggplot2 package

developed by Hadley Wickham), import/export capabilities, reporting tools (knitr, Sweave), etc.

本句中，which include specialized statistical techniques, graphical devices (such as the ggplot2 package developed by Hadley Wickham), import/export capabilities, reporting tools (knitr, Sweave), etc.是一个非限定性定语从句，对 user-created packages 进行补充说明。developed by Hadley Wickham 是一个过去分词短语，作定语，修饰和限定 ggplot2 package。

[5] The "Task Views" page (subject list) on the CRAN website lists a wide range of tasks (in fields such as Finance, Genetics, High Performance Computing, Machine Learning, Medical Imaging, Social Sciences and Spatial Statistics) to which R has been applied and for which packages are available.

本句中，on the CRAN website 是一个介词短语，作定语，修饰和限定 The "Task Views" page。to which R has been applied and for which packages are available 是两个介词前置的定语从句，修饰和限定 a wide range of tasks。

[6] Other R package resources include Crantastic, a community site for rating and reviewing all CRAN packages, and R-Forge, a central platform for the collaborative development of R packages, R-related software, and projects.

本句中，a community site for rating and reviewing all CRAN packages 对 Crantastic packages 进行补充说明。a central platform for the collaborative development of R packages, R-related software, and projects 对 user-created packages 进行补充说明。

Exercises

【Ex. 1】 根据课文内容回答问题。

1. What is R?
2. What is the purpose for statisticians and data miners widely use the R language?
3. In what languages is the source code for the R software environment written primarily?
4. By whom was R created? And where?
5. What are the statistical and graphical techniques R and its libraries implement?
6. Why does R has has stronger object-oriented programming facilities than most statistical computing languages?
7. What do R's data structures include?
8. What do R's extensible object system include?
9. What has R also been identified by the FDA as?
10. What do text editors and integrated development environments (IDEs) with some support for R include?

【Ex. 2】 中译英或者英译中。

1. *n.*项目，方案，工程   1. _____
2. *adj.*标准的，普遍的 *n.*标准   2. _____
3. *vt.* & *n.*支持，支撑   3. _____
4. *n.*回归   4. _____
5. *adj.*不变的，照旧的   5. _____
6. website   6. _____
7. benchmark   7. _____
8. clustering   8. _____
9. development   9. _____
10. interface   10. _____

【Ex. 3】 短文翻译。

What is R?

1. Introduction to R

R is a language and environment for statistical computing and graphics. It is a GNU project which is similar to the S language and environment which was developed at Bell Laboratories (formerly AT&T, now Lucent Technologies) by John Chambers and colleagues. R can be considered as a different implementation of S. There are some important differences, but much code written for S runs unaltered under R.

R provides a wide variety of statistical (linear and nonlinear modelling, classical statistical tests, time-series analysis, classification, clustering, …) and graphical techniques, and is highly extensible. The S language is often the vehicle of choice for research in statistical methodology, and R provides an Open Source route to participation in that activity.

One of R's strengths is the ease with which well-designed publication-quality plots can be produced, including mathematical symbols and formulae where needed.

R is available as Free Software under the terms of the Free Software Foundation's GNU General Public License in source code form. It compiles and runs on a wide variety of UNIX platforms and similar systems (including FreeBSD and Linux), Windows and MacOS.

2. The R environment

R is an integrated suite of software facilities for data manipulation, calculation and graphical display. It includes:

- An effective data handling and storage facility,
- A suite of operators for calculations on arrays, in particular matrices,
- A large, coherent, integrated collection of intermediate tools for data analysis,
- Graphical facilities for data analysis and display either on-screen or on hardcopy, and
- A well-developed, simple and effective programming language which includes conditionals, loops, user-defined recursive functions and input and output facilities.

R, like S, is designed around a true computer language, and it allows users to add additional functionality by defining new functions. For computationally-intensive tasks, C, C++ and Fortran code can be linked and called at run time. Advanced users can write C code to manipulate R objects directly.

【Ex. 4】 将下列词填入适当的位置（每词只用一次）。

| readable | type | code | interpreted | compiled |
| operations | program | save | processing | libraries |

What Is Python?

1. What Is Python?

The Python programming language is freely available and makes solving a computer problem almost as easy as writing out your thoughts about the solution. The __(1)__ can be written once and run on almost any computer without needing to change the __(2)__.

2. How Python Is Used?

Python is a general purpose programming language that can be used on any modern computer operating system. It can be used for __(3)__ text, numbers, images, scientific data and just about anything else you might __(4)__ on a computer. It is used daily in the __(5)__ of the Google search engine, the video-sharing website YouTube, NASA and the New York Stock Exchange. These are but a few of the places where Python plays important roles in the success of business, government and non-profit organizations; there are many others.

Python is an __(6)__ language. This means that it is not converted to computer-readable code before the program is run but at runtime. In the past, this __(7)__ of language was called a scripting language, intimating its use was for trivial tasks. However, programming languages such as Python have forced a change in that nomenclature. Increasingly, large applications are written almost exclusively in Python.

3. How does Python compare to Java?

Both Python and Java are object-oriented languages with substantial __(8)__ of pre-written code that can be run on almost any operating system. However, their implementations are vastly different.

Java is neither an interpreted language nor a compiled language. It is a bit of both. When compiled, Java programs are compiled to bytecode—a Java-specific type of code. When the program is run, this bytecode is run through a Java Runtime Environment to convert it to machine code, which is __(9)__ and executable by the computer. Once compiled to bytecode, Java programs cannot be modified.

Python programs, on the other hand, are typically __(10)__ at the time of running, when the Python interpreter reads the program. However, they can be compiled to computer-readable machine code. Python does not use an intermediary step for platform independence. Instead, platform independence is in the implementation of the interpreter.

Text B

Python Programming Language

Python is a widely used high-level, general-purpose, interpreted, dynamic programming language. Its design philosophy emphasizes code readability, and its syntax allows programmers to express concepts in fewer lines of code than possible in languages such as C++ or Java. The language provides constructs intended to enable clear programs on both a small and large scale.

Python supports multiple programming paradigms, including object-oriented, imperative and functional programming or procedural styles. It features a dynamic type system and automatic memory management and has a large and comprehensive standard library.

Python interpreters are available for many operating systems, allowing Python code to run on a wide variety of systems. Using third-party tools, such as Py2exe or Pyinstaller, Python code can be packaged into stand-alone executable programs for some of the most popular operating systems, so Python-based software can be distributed to, and used on, those environments with no need to install a Python interpreter.

1. Features and philosophy

Python is a multi-paradigm programming language: object-oriented programming and structured programming are fully supported, and many language features support functional programming and aspect-oriented programming. Many other paradigms are supported via extensions, including design by contract and logic programming.

Rather than requiring all desired functionality to be built into the language's core, Python was designed to be highly extensible. Python can also be embedded in existing applications that need a programmable interface.

2. Syntax and semantics

Python is intended to be a highly readable language. It is designed to have an uncluttered visual layout, often using English keywords where other languages use punctuation. Furthermore, Python has fewer syntactic exceptions and special cases than C or Pascal.

2.1 Indentation

Python uses whitespace indentation, rather than curly braces or keywords, to delimit blocks; this feature is also termed the off-side rule. An increase in indentation comes after certain statements; a decrease in indentation signifies the end of the current block.

2.2 Statements and control flow

Python's statements include (among others):
- The assignment statement (token "=", the equals sign). This operates differently than in traditional imperative programming languages, and this fundamental mechanism (including the nature of Python's version of *variables*) illuminates many other features of the language. Assignment in C, e.g., x = 2, translates to "typed variable name x receives a copy of numeric value 2". The (right-hand) value is copied into an allocated storage location for which the (left-hand) variable name is the symbolic address. The memory allocated to the variable is large enough (potentially quite large) for the declared type. In the simplest case of Python assignment, using the same example, x = 2, translates to "(generic) name x receives a reference to a separate, dynamically allocated object of numeric (int) type of value 2." This is termed *binding* the name to the object. Since the name's storage location doesn't *contain* the indicated value, it is improper to call it a *variable*. Names may be subsequently rebound at any time to objects of greatly varying types, including strings, procedures, complex objects with data and methods, etc. Successive assignments of a common value to

multiple names, e.g., x = 2; y = 2; z = 2 result in allocating storage to (at most) three names and one numeric object, to which all three names are bound. Since a name is a generic reference holder it is unreasonable to associate a fixed data type with it. However at a given time a name will be bound to some object, which will have a type; thus there is dynamic typing.

- The if statement, which conditionally executes a block of code, along with else and elif (a contraction of else-if).
- The for statement, which iterates over an iterable object, capturing each element to a local variable for use by the attached block.
- The while statement, which executes a block of code as long as its condition is true.
- The try statement, which allows exceptions raised in its attached code block to be caught and handled by except clauses; it also ensures that clean-up code in a finally block will always be run regardless of how the block exits.
- The class statement, which executes a block of code and attaches its local namespace to a class, for use in object-oriented programming.
- The def statement, which defines a function or method.
- The with statement (from Python 2.5), which encloses a code block within a context manager (for example, acquiring a lock before the block of code is run and releasing the lock afterwards, or opening a file and then closing it), allowing Resource Acquisition Is Initialization (RAII)-like behavior.
- The pass statement, which serves as a NOP. It is syntactically needed to create an empty code block.
- The assert statement, used during debugging to check for conditions that ought to apply.
- The yield statement, which returns a value from a generator function. From Python 2.5, yield is also an operator. This form is used to implement coroutines.
- The import statement, which is used to import modules whose functions or variables can be used in the current program.
- The print statement was changed to the print( ) function in Python 3.

2.3 Expressions

Some Python expressions are similar to languages such as C and Java, while some are not:

- Addition, subtraction, and multiplication are the same, but the behavior of division differs. Python also added the ** operator for exponentiation.

- As of Python 3.5, it supports matrix multiplication directly with the @ operator, versus C and Java, which implement these as library functions. Earlier versions of Python also used methods instead of an infix operator.
- In Python, == compares by value, versus Java, which compares numerics by value and objects by reference. (Value comparisons in Java on objects can be performed with the equals() method.) Python's is operator may be used to compare object identities (comparison by reference). In Python, comparisons may be chained, for example a <= b <= c.
- Python uses the words and, or, not for its boolean operators rather than the symbolic &&, ||, ! used in Java and C.
- Python has a type of expression termed a *list comprehension*. Python 2.4 extended list comprehensions into a more general expression termed a *generator expression*.
- Anonymous functions are implemented using lambda expressions; however, these are limited in that the body can only be one expression.
- Conditional expressions in Python are written as x if c else y (different in order of operands from the c ? x : y operator common to many other languages).
- Python makes a distinction between lists and tuples. Lists are written as [1, 2, 3], are mutable, and cannot be used as the keys of dictionaries (dictionary keys must be immutable in Python). Tuples are written as (1, 2, 3), are immutable and thus can be used as the keys of dictionaries, provided all elements of the tuple are immutable. The parentheses around the tuple are optional in some contexts. Tuples can appear on the left side of an equal sign; hence a statement like x, y = y, x can be used to swap two variables.
- Python has a "string format" operator %. This functions analogous to printf format strings in C.
- Python has various kinds of string literals:

    (1) Strings delimited by single or double quote marks. Unlike in Unix shells, Perl and Perl-influenced languages, single quote marks and double quote marks function identically. Both kinds of string use the backslash (\) as an escape character and there is no implicit string interpolation such as "$spam".

    (2) Triple-quoted strings, which begin and end with a series of three single or double quote marks. They may span multiple lines and function like here documents in shells, Perl and Ruby.

    (3) Raw string varieties, denoted by prefixing the string literal with an r. No escape sequences are interpreted; hence raw strings are useful where literal backslashes are common, such as regular expressions and Windows-style paths. Compare "@-quoting" in C#.

- Python has array index and array slicing expressions on lists, denoted as a[key], a[start:stop] or a[start:stop:step]. Indexes are zero-based, and negative indexes are relative to the end. Slices take elements from the *start* index up to, but not including, the *stop* index. The third slice parameter, called *step* or *stride*, allows elements to be skipped and reversed. Slice indexes may be omitted, for example a[:] returns a copy of the entire list. Each element of a slice is a shallow copy.

In Python, a distinction between expressions and statements is rigidly enforced, in contrast to languages such as Common Lisp, Scheme, or Ruby. This leads to duplicating some functionality.

Statements cannot be a part of an expression, so list and other comprehensions or lambda expressions, all being expressions, cannot contain statements. A particular case of this is that an assignment statement such as a = 1 cannot form part of the conditional expression of a conditional statement. This has the advantage of avoiding a classic C error of mistaking an assignment operator = for an equality operator == in conditions: if (c = 1) { ... } is valid C code but if c = 1: ... causes a syntax error in Python.

2.4 Mathematics

Python has the usual C arithmetic operators (+,−, *, /, %). It also has ** for exponentiation, e.g. 5**3 == 125 and 9**0.5 == 3.0, and a new matrix multiply @ operator is included in version 3.5.

Python provides a round function for rounding a float to the nearest integer. For tie-breaking, versions before 3 use round-away-from-zero: round(0.5) is 1.0, round(−0.5) is −1.0. Python 3 uses round to even: round(1.5) is 2, round(2.5) is 2.

Python allows boolean expressions with multiple equality relations in a manner that is consistent with general use in mathematics. For example, the expression a < b < c tests whether a is less than b and b is less than c. C-derived languages interpret this expression differently: in C, the expression would first evaluate a < b, resulting in 0 or 1, and that result would then be compared with c.

Due to Python's extensive mathematics library, it is frequently used as a scientific scripting language to aid in problems such as numerical data processing and manipulation.

3. Libraries

Python has a large standard library, commonly cited as one of Python's greatest strengths, providing tools suited to many tasks. For Internet-facing applications, many standard formats and protocols (such as MIME and HTTP) are supported. Modules for creating graphical user interfaces, connecting to relational databases, pseudorandom number generators, arithmetic with arbitrary precision decimals, manipulating regular expressions, and doing unit testing are also included.

The standard library is not needed to run Python or embed it in an application. For example, Blender 2.49 omits most of the standard library.

As of January 2016, the Python Package Index, the official repository of third-party software for Python, contains more than 72000 packages offering a wide range of functionality, including:

- graphical user interfaces, web frameworks, multimedia, databases, networking and communications
- test frameworks, automation and web scraping, documentation tools, system administration
- scientific computing, text processing, image processing

4. Development environments

Most Python implementations can function as a command line interpreter, for which the user enters statements sequentially and receives the results immediately (read–eval–print loop (REPL)). In short, Python acts as a command-line interface or shell.

Other shells add abilities beyond those in the basic interpreter, including IDLE and IPython. While generally following the visual style of the Python shell, they implement features like auto-completion, session state retention, and syntax highlighting.

In addition to standard desktop integrated development environments (Python IDEs), there are also web browser-based IDEs, Sage (intended for developing science and math-related Python programs), and a browser-based IDE and hosting environment, PythonAnywhere.

New Words

high-level	['hai-'levəl]	adj.高级的
general-purpose	['dʒenərəl'pə:pəs]	adj.多种用途的
philosophy	[fi'lɔsəfi]	n.哲学，哲学体系
emphasize	['emfəsaiz]	vt.强调，着重
readability	[,ri:də'biliti]	n.易读，可读性
programmer	['prəugræmə]	n.程序员
imperative	[im'perətiv]	n.命令
		adj.命令的
unclutter	[ˌʌn'klʌtə]	vt.使整洁，整理
exception	[ik'sepʃən]	n.异常，例外

indentation	[ˌinden'teiʃən]	n.缩排
keyword	['ki:wəd]	n.关键字
statement	['steitmənt]	n.语句
declared	[di'klɛəd]	adj.声明的
improper	[im'prɔpə]	adj.不适当的，不合适的，不正确的
string	[striŋ]	n.串
successive	[sək'sesiv]	adj.连续的
contraction	[kən'trækʃən]	n.缩写式，紧缩
iterate	['itəreit]	vt.重复
clause	[klɔ:z]	n.子句
namespace	['neimspeis]	n.名空间
class	[klɑ:s]	n.类
enclose	[in'kləuz]	vt.封装
manager	['mænidʒə]	n.管理器
debugging	[di:'bʌgiŋ]	n.调试
generator	['dʒenəreitə]	n.生成器
coroutine	[ˌkəru:'ti:n]	n.协同程序
expression	[iks'preʃən]	n.表达式
exponentiation	['ekspəu,nenʃi'eiʃən]	n.求幂
operator	['ɔpəreitə]	n.运算符
infix	[in'fiks]	n.中缀
		vt.让……插进
mutable	['mju:təbl]	adj.可变的，易变的
immutable	[i'mju:təbl]	adj.不可变的，不能变的
tuple	['tʌpl]	n.元组
parentheses	[pə'renθəsi:z]	n.圆括号
optional	['ɔpʃənəl]	adj.可选择的
context	['kɔntekst]	n.上下文，情景
analogous	[ə'næləgəs]	adj.类似的，相似的
literal	['litərəl]	adj.文字的，照字面上的
backslash	['bækslæʃ]	n.反斜线符号
prefix	['pri:fiks]	n.前缀
parameter	[pə'ræmitə]	n.参数，参量
reverse	[ri'və:s]	v.翻转，颠倒
omit	[əu'mit]	vt.省略，遗漏
rounding	['raundiŋ]	n.舍入，取整
module	['mɔdju:l]	n.模块

pseudorandom	[ˌpsjuːdəʊˈrændəm]	adj.伪随机的
multimedia	[ˌmʌltiˈmiːdiə]	n.多媒体
communication	[kəˌmjuːniˈkeiʃn]	n.通信

📖 Phrases

dynamic programming language	动态编程语言
design philosophy	设计原理
lines of code	代码行
memory management	内存管理
standard library	标准库
third-party tool	第三方工具
structured programming	结构化编程
aspect-oriented programming	面向切面编程
design by contract	契约设计
logic programming	逻辑编程
special case	特殊情况
curly braces	大括号，花括号
off-side rule	越位规则
control flow	控制流
assignment statement	赋值语句
symbolic address	符号地址
storage location	存储位置，存储单元
matrix multiplication	矩阵乘法
library function	库函数
object identity	对象标识
boolean operator	布尔运算符，逻辑运算符
list comprehension	列表解析，列表推导
generator expression	生成器表达式
anonymous function	匿名函数
lambda expression	λ表达式
conditional expression	条件表达式
make a distinction between…	对……加以区别
string literal	字符串字面量
single quote mark	单引号
double quote mark	双引号
escape character	转义字符

string interpolation	字符串插值
triple-quoted string	三重引号字符串
regular expression	正则表达式
array index	数组下标
array slicing	数组切片
shallow copy	浅拷贝
particular case	特别情况，特例
have the advantage of	胜过
boolean expression	逻辑表达式，布尔表达式
in a manner	在某种意义上
pseudorandom number generator	伪随机数产生器
web scraping	网络爬虫，网络数据抓取
image processing	图像加工，图像处理
development environment	开发环境
command line interpreter	命令行解释程序
session state retention	会话状态保留
hosting environment	托管环境

Abbreviations

RAII (Resource Acquisition Is Initialization)	资源获得即初始化
NOP (No Operation)	无操作
MIME (Multipurpose Internet Mail Extensions)	多用途互联网邮件扩展
HTTP（HyperText Transfer Protocol）	超文本传输协议
REPL (Read-Eval-Print Loop)	读取-求值-打印循环

Exercises

【Ex. 5】 根据课文内容回答问题。

1. What is Python?
2. What are the key features of Python?
3. What does the if statement conditionally do?
4. What does the try statement do?
5. What does Python 3.5 support matrix multiplication with?
6. What distinction does Python make between lists and tuples?
7. What do triple-quoted strings begin and end with?
8. What is the third slice parameter called? What does it do?

9. What is one of Python's greatest strengths?
10. What can most Python implementations function as?

参考译文

R 编程语言

R 语言是用于统计计算和图形的编程语言和软件环境，由统计计算 R 基金会支持。统计学家和数据挖掘者广泛使用 R 语言来开发统计软件和分析数据。民意测验、对数据挖掘者的调查以及对学术文献数据库研究都显示 R 语言近年来的受欢迎程度大大提高。

R 语言是一个 GNU 项目。用于 R 软件环境的源代码主要用 C 语言、Fortran 语言和 R 语言来编写。R 语言可以持 GNU 通用公共许可证免费获得，也提供用于各种操作系统的预编译二进制版本。虽然 R 语言有一个命令行界面，但也有几个图形前端可供使用。

1. 历史

R 语言是 S 语言与词汇作用域语义的结合。S 语言由 John Chambers 在贝尔实验室创建。虽然两者有一些重要的区别，但为 S 编写的大部分代码无须修改即可运行。

R 语言由新西兰奥克兰大学的罗斯·伊哈卡（Ross Ihaka）和罗伯特·杰特曼（Robert Gentleman）创建，目前由钱伯斯（Chambers）所在的 R 语言开发核心团队开发。该项目于 1992 年构思，最初版本于 1994 年发布，2000 年发布了稳定的 beta 版本。

2. 统计功能

R 语言及其库采用了各种统计和图形技术，包括线性和非线性建模、经典统计测试、时间序列分析、分类及聚类等。R 语言可以通过函数和扩展部件轻松扩充，R 社区在软件包方面的积极贡献值得注意。R 语言的许多标准函数都是用 R 语言编写的，这使得用户可以轻松地实现所选择算法。对于计算量大的任务，可以在运行时链接和调用 C、C++ 和 Fortran 代码。高级用户可以编写 C、C++、Java、.NET 或 Python 代码来直接处理 R 对象。R 语言使用用户提交的软件包极大地扩展了其应用，这些软件包可用于特定函数或特定的研究领域。由于其继承了 S 语言，因而当采用面向对象编程技术时，R 语言比大多数统计计算语言更便利。可以通过词法作用域规则轻松扩展 R 语言。

R 语言的另一个优点是静态图形处理能力强，可以生成出版品质的图形，包括数学符号。使用其他软件包也可以处理动态和交互式图形。

R 语言有其自身的 LaTeX 类文档格式 Rd，广泛支持各类文档，既支持多种在线格式，也支持硬拷贝。

3. 编程功能

R 语言是一种解释语言，用户通常通过命令行解释器访问它。如果用户在 R 命令提示符下键入"2＋2"并按回车键，则计算机将回复 4。

像其他类似的语言（如 APL 和 MATLAB）一样，R 语言支持矩阵运算。R 语言的数据结构包括向量、矩阵、数组、数据帧（类似于关系数据库中的表）和列表。R 语言的可扩展对象系统包括用于回归模型、时间序列和地理空间坐标（以及其他）的对象。标量数据类型从不是 R 语言的数据结构。相反，标量被表示为长度为 1 的向量。

R 语言支持带函数的过程化编程，对于一些函数，也可用于带类函数的面向对象编程。类函数根据给其传递的参数类别而有所不同。换句话说，类函数为特定的类对象指定特定的函数（或方法）。例如，R 语言具有通用 print 功能，可以使用简单的 print (objectname) 语法在 R 语言中打印几乎每一类对象。

虽然 R 语言主要由统计学家和其他从业者用于统计计算和软件开发的环境，但也可以用作一般矩阵计算工具箱，其性能基准与 GNU Octave 或 MATLAB 相当。

4. 软件包

可通过用户创建的软件包来扩展 R 语言的性能，这些软件包专用于统计技术、图形设备（如 Hadley Wickham 开发的 ggplot2 包）、导入/导出功能、报告工具（knitr、Sweave）等方面。这些软件包主要用 R 语言开发，有时也用 Java、C、C++和 Fortran 来开发。

R 语言的安装包括一套核心套件，还有超过 7801 个附加软件包（截至 2016 年 1 月），这些附加软件包用于 CRAN（R 综合归档网）、Bioconductor（生物导体）、Omegahat、GitHub 和其他软件库。

CRAN 网站上的"任务视图"页面（主题列表）列出了可用 R 语言完成的各种任务（诸如金融、遗传学、高性能计算、机器学习、医学影像、社会科学和空间统计学等领域）和可用的软件包。R 语言已经被 FDA 认定为适合解释临床研究数据。

其他 R 语言包资源包括 Crantastic（用于评估和审查所有 CRAN 软件包的社区网站）以及 R-Forge（这是 R 语言软件包以及 R 语言相关软件和项目协同开发的中心平台）。R-Forge 还提供许多未发布的 beta 测试版软件包和 CRAN 软件包的开发版本。

5. 界面

5.1 图形用户界面

- Architect——基于 Eclipse 和 StatET 的数据科学的跨平台开源 IDE。
- DataJoy——online R Editor 专注于数据科学与协作的初学者。
- Deducer——用于菜单驱动数据分析的 GUI（类似于 SPSS / JMP / Minitab）。

- Java GUI for R——跨平台的独立 R 终端和基于 Java（也称为 JGR）的编辑器。
- Number Analytics——基于云的业务分析（类似于 SPSS）的 GUI。
- Rattle GUI——基于 RGtk2 的跨平台 GUI，专为数据挖掘而设计。
- R Commander——基于 tcltk 的跨平台菜单驱动的 GUI（也可以使用几个 Rcmdr 插件）。
- Revolution R Productivity Environment (RPE)——提供 Revolution Analytics 的基于 Visual Studio 的 IDE，并具有基于网页的页面操作界面。
- RGUI——带有用于 Microsoft Windows 的预编译版本。
- RKWard——用于 R 的可扩展 GUI 和 IDE。
- RStudio——跨平台开源 IDE（也可以在远程 Linux 服务器上运行）。

5.2 编辑器和 IDE

支持 R 语言的文本编辑器和集成开发环境（IDE）包括 ConTEXT、Eclipse（StatET）、Emacs（Emacs Speaks Statistics）、LyX（用于 knitr 和 Sweave 的模块）、Vim、jEdit、Kate、RStudio、Sublime Text、TextMate、Atom、WinEdt（R Package RWinEdt）、Tinn-R、Notepad ++和 Architect。

5.3 脚本语言

可以从几种脚本语言（如 Python、Perl、Ruby、F#和 Julia）访问 R 功能。R 本身的脚本可以通过一个名叫"小可爱"（littler）的前端访问。

6. 与 SAS、SPSS 和 Stata 的比较

一个共识是：与其他受欢迎的统计软件包（如 SAS、SPSS 和 Stata）相比，R 语言更优异。对统计软件的所有基本特色功能进行比较后，人们认为 R 是最好的统计软件。

2015 年 1 月，《纽约时报》刊登的一篇文章指出：R 语言在数据分析师中获得广泛认可，并对 SAS 等商业统计软件所占据的市场份额构成了威胁。

Unit 5

Text A

Data Structure

A data structure is a specialized format for organizing and storing data. General data structure types include the array, the file, the record, the table, the tree, and so on. Any data structure is designed to organize data to suit a specific purpose so that it can be accessed and worked with in appropriate ways. In computer programming, a data structure may be selected or designed to store data for the purpose of working on it with various algorithms.

1. Array

(1) In data storage, an array is a method for storing information on multiple devices.

(2) In general, an array is a number of items arranged in some specified way, for example, in a list or in a three-dimensional table.

(3) In computer programming languages, an array is a group of objects with the same attributes that can be addressed individually, using such techniques as subscripting.

(4) In random access memory (RAM), an array is the arrangement of memory cells.

2. File

(1) In data processing, a file is a related collection of records. For example, you might put the records you have on each of your customers in a file. In turn, each record would consist of fields for individual data items, such as customer name, customer number, customer

address, and so forth. By providing the same information in the same fields in each record so that all records are consistent, your file will be easily accessible for analysis and manipulation by a computer program. This use of the term has become somewhat less important with the advent of the database and its emphasis on the table as a way of collecting record and field data. In mainframe systems, the term data set is generally synonymous with file but implies a specific form of organization recognized by a particular access method. Depending on the operating system, files (and data sets) are contained within a catalog, directory, or folder.

(2) In any computer system, especially in personal computers, a file is an entity of data available to system users (including the system itself and its application programs) that is capable of being manipulated as an entity (for example, moved from one file directory to another). The file must have a unique name within its own directory. Some operating systems and applications describe files with given formats by giving them a particular file name suffix. The file name suffix is also known as a file name extension. For example, a program or executable file is sometimes given or required to have an ".exe" suffix. In general, the suffixes tend to be as descriptive of the formats as they can within the limits of the number of characters allowed for suffixes by the operating system.

3. Record

(1) In computer data processing, a record is a collection of data items arranged for processing by a program. Multiple records are contained in a file or data set. The organization of data in the record is usually prescribed by the programming language that defines the record's organization and/or by the application that processes it. Typically, records can be of fixed-length or be of variable length with the length information contained within the record.

(2) In a database, a record, sometimes called a row, is a group of fields within a table that are relevant to a specific entity. For example, in a table called customer contact information, a row would likely contain fields such as: ID number, name, street address, city, telephone number and so on.

4. Table

In computer programming, a table is a data structure used to organize information, just as it is on paper. There are many different types of computer-related tables, which work in a number of different ways. The following are examples of the more common types.

(1) In data processing, a table, also called an array, is an organized grouping of fields. Tables may store relatively permanent data, or may be frequently updated. For example, a table contained in a disk volume is updated when sectors are being written.

(2) In a relational database, a table, sometimes called a file, organizes the information about a single topic into rows and columns. For example, a database for a business would typically contain a table for customer information, which would store customers' account numbers, addresses, phone numbers, and so on as a series of columns. Each single piece of data, such as the account number, is a field in the table. A column consists of all the entries in a single field, such as the telephone numbers of all the customers. Fields, in turn, are organized as records, which are complete sets of information, such as the set of information about a particular customer, each of which comprises a row. The process of normalization determines how data will be most effectively organized into tables.

(3) A decision table, often called a truth table, which can be computer-based or simply drawn up on paper, contains a list of decisions and the criteria on which they are based. All possible situations for decisions should be listed, and the action to take in each situation should be specified. A rudimentary example: For a traffic intersection, the decision to proceed might be expressed as yes or no and the criteria might be the light is red or the light is green.

A decision table can be inserted into a computer program to direct its processing according to decisions made in different situations. Changes to the decision table are reflected in the program.

(4) An HTML table is used to organize Web page elements spatially or to create a structure for data that is best displayed in tabular form, such as lists or specifications.

New Words

specialized	[ˈspeʃəlaizd]	adj.专用的，专门的
organize	[ˈɔːgənaiz]	vt.组织；构成，组成
array	[əˈrei]	n.数组，排列
record	[ˈrekɔːd]	n.记录
	[riiˈkɔːd]	vt.记录；录音
table	[ˈteibl]	n.表，表格
appropriate	[əˈprəupriət]	adj.正确的，适当的
various	[ˈvɛəriəs]	adj.不同的，各种各样的，多方面的，多样的
subscript	[ˈsʌbskript]	adj.下标
collection	[kəˈlekʃən]	n.集合，收集来的总和
item	[ˈaitəm]	n.项目
consistent	[kənˈsistənt]	adj.一致的，调和的，相容的
accessible	[əkˈsesəbl]	adj.易接近的，可访问的，易受影响的
manipulation	[məˌnipjuˈleiʃən]	n.处理，操作
advent	[ˈædvənt]	n.出现，到来

emphasis	['emfəsis]	n.强调，重点
imply	[im'plai]	vt.暗示，意味
suffix	['sʌfiks]	n.后缀；下标
prescribe	[pris'kraib]	v.指示，规定
define	[di'fain]	vt.定义，详细说明
length	[leŋθ]	n.长度
row	[rau]	n.行，排
relevant	['relivənt]	adj.有关的，相应的
common	['kɔmən]	adj.共同的，公共的，公有的，普通的
permanent	['pə:mənənt]	adj.永久的，持久的
frequently	['fri:kwəntli]	adv.常常，频繁地
sector	['sektə]	n.扇区
normalization	[ˌnɔ:məlai'zeiʃən]	n.规范化，正常化，标准化
criteria	['krai'tiəriə]	n.标准
rudimentary	[ru:di'mentəri]	adj.基本的，初步的
intersection	[ˌintə:'sekʃən]	n.交集，十字路口，交叉点
spatial	['speiʃəl]	adj.空间的，立体的，三维的

Phrases

data structure	数据结构
and so on	等等
a number of	许多的
three-dimensional table	三维表
memory cell	内存单元
and so forth	等等
data set	数据集
file name extension	文件扩展名
account number	账号
in turn	依次，轮流
decision table	判定表，决策表
truth table	真值表
draw up	草拟

Abbreviations

ID（Identification, Identity）	身份

✎ Notes

[1] Any data structure is designed to organize data to suit a specific purpose so that it can be accessed and worked with in appropriate ways.

本句中，to organize data to suit a specific purpose so that it can be accessed and worked with in appropriate ways 是一个动词不定式短语，作目的状语，修饰 is designed。在该短语中，to suit a specific purpose 也是一个动词不定式短语，作目的状语，修饰 to organize，so that it can be accessed and worked with in appropriate ways 是一个目的状语从句。

[2] In computer programming languages, an array is a group of objects with the same attributes that can be addressed individually, using such techniques as subscripting.

本句中，with the same attributes 是一个介词短语，作定语，修饰和限定 a group of objects。that can be addressed individually 是一个定语从句，也修饰和限定 a group of objects，using such techniques as subscripting 是一个现在分词短语，作方式状语，修饰从句的谓语 can be addressed。

[3] By providing the same information in the same fields in each record so that all records are consistent, your file will be easily accessible for analysis and manipulation by a computer program.

本句中，By providing the same information in the same fields in each record so that all records are consistent 是一个现在分词短语，作方式状语，修饰谓语 will be easily accessible。在该短语中，so that all records are consistent 是一个结果状语从句，修饰谓语 providing。

英语中，so that 既可以引导一个目的状语从句，也可以引导一个结果状语从句。请看下例：

We asked the professor to speak louder so that we could hear him.

我们请教授讲话声再大一些，以便让我们能听清。（目的状语从句）

Mary didn't plan her time well, so that she didn't finish the work in time.

玛丽没有把时间计划好，结果没有按时完成这项工作。（结果状语从句）

[4] In any computer system but especially in personal computers, a file is an entity of data available to system users (including the system itself and its application programs) that is capable of being manipulated as an entity (for example, moved from one file directory to another).

本句中，available to system users 是一个现在分词短语，作定语，修饰和限定 an entity of data。that is capable of being manipulated as an entity 是一个定语从句，也修饰和限定 an entity of data。

[5] The organization of data in the record is usually prescribed by the programming language

that defines the record's organization and/or by the application that processes it.

本句中，and/or 连接了 by 引导的两个方式状语。that defines the record's organization 是一个定语从句，修饰和限定 the programming language。

[6] For example, a database for a business would typically contain a table for customer information, which would store customers' account numbers, addresses, phone numbers, and so on as a series of columns.

本句中，which would store customers' account numbers, addresses, phone numbers, and so on as a series of columns 是一个非限定性定语从句，对 a table for customer information 进行补充说明。

[7] Fields, in turn, are organized as records, which are complete sets of information, such as the set of information about a particular customer, each of which comprises a row.

本句中，which are complete sets of information 是一个非限定性定语从句，对 records 进行补充说明。such as the set of information about a particular customer 是对 complete sets of information 的举例说明，each of which comprises a row 是一个非限定性定语从句，对 the set of information about a particular customer 进行补充说明。

英语中，定语从句还可以由名词（代词/数词）+ of + which（whom）来引导，表示部分与整体的关系。注意不要误用 which 和 whom。which 指物，whom 用来指人。请看下例：

Peter's father knows a lot of people, many of whom are professors.

彼得的爸爸认识许多人，其中许多是教授。

She bought many books yesterday, five of which are on ERP.

她昨天买了许多书，其中 5 本是 ERP 方面的。

Exercises

【Ex. 1】根据课文内容回答问题。

1. What is a data structure?
2. What is an array in computer programming languages?
3. Must the file have a unique name within its own directory?
4. How do some operating systems and applications describe files with given formats?
5. How is the organization of data in the record usually prescribed?
6. What is a table in computer programming?
7. What is a table in data processing?
8. What is a table in a relational database?
9. What is a decision table often called? What does it contain?
10. What is an HTML table used to do?

【Ex. 2】根据下面的英文解释，写出相应的英文词汇。

1. _____: A signal to a computer that stops the execution of a running program so that another action can be performed.
2. _____: A collection of related, often adjacent items of data, treated as a unit.
3. _____: In word processing, a block of text formatted in aligned rows and columns.
4. _____: A multi-element data structure that has a linear organization but that allows elements to be added or removed in any order.
5. _____: A distinguishing character or symbol written directly beneath or next to and slightly below a letter or number.
6. _____: An affix added to the end of a word or stem.
7. _____: To make or write a definition.
8. _____: A series of objects placed next to each other, usually in a straight line.
9. _____: A bit or a set of bits on a magnetic storage device making up the smallest addressable unit of information.
10. _____: To organize data, typically a set of records, in a particular order.

【Ex. 3】把下列句子翻译为中文。

1. Star topologies are normally implemented using twisted pair cable, specifically unshielded twisted pair (UTP).
2. A video card is the part of your computer that transforms video data into the visual display you see on your monitor.
3. A multi-user operating system allows many different users to take advantage of the computer's resources simultaneously.
4. Address is the unique location of an information site on the Internet, a specific file (for example, a Web page), or an E-mail user.
5. Over the years, ARPA has funded many projects in computer science research, many of which had a profound effect on the state of the art.
6. In truth of course by making the creation of more complex software practical, computer languages have merely created new types of software bugs.
7. A computer virus is a program designed to spread itself by first infecting executable files or the system areas of hard and floppy disks and then making copies of itself.
8. When the entire RAM is being used (for example if there are many programs open at the same time) the computer will swap data to the hard drive and back to give the impression that there is slightly more memory.
9. The compiler ignores all comments.
10. You can E-mail your document without ever leaving word.

【Ex. 4】将下列词填入适当的位置（每词只用一次）。

| records | leaves | beyond | random | database |
| two | power | depth | nodes | end |

A binary tree is a method of placing and locating files (called records or keys) in a __(1)__, especially when all the data is known to be in __(2)__ access memory (RAM). The algorithm finds data by repeatedly dividing the number of ultimately accessible __(3)__ in half until only one remains.

In a tree, records are stored in locations called __(4)__. This name derives from the fact that records always exist at __(5)__ points; there is nothing __(6)__ them. Branch points are called __(7)__. The order of a tree is the number of branches (called children) per node. In a binary tree, there are always __(8)__ children per node, so the order is 2. The number of leaves in a binary tree is always a __(9)__ of 2. The number of access operations required to reach the desired record is called the __(10)__ of the tree.

Text B

Structured Data, Semi-structured Data and Unstructured Data

1. Structured Data

Structured data refers to any data that resides in a fixed field within a record or file. This includes data contained in relational databases and spreadsheets.

1.1 Characteristics of Structured Data

Structured data first depends on creating a data model – a model of the types of business data that will be recorded and how they will be stored, processed and accessed. This includes defining what fields of data will be stored and how that data will be stored: data type (numeric, currency, alphabetic, name, date, address) and any restrictions on the data input (number of characters; restricted to certain terms such as Mr., Ms. or Dr.; M or F).

Structured data has the advantage of being easily entered, stored, queried and analyzed. At one time, because of the high cost and performance limitations of storage, memory and processing, relational databases and spreadsheets using structured data were the only ways to effectively manage data. Anything that couldn't fit into a tightly organized structure would have to be stored on paper in a filing cabinet.

1.2 Managing Structured Data

Structured data is often managed using Structured Query Language (SQL) – a programming language created for managing and querying data in relational database management systems. Originally developed by IBM in the early 1970s and later developed commercially by Relational Software, Inc. (now Oracle Corporation).

Structured data was a huge improvement over strictly paper-based unstructured systems, but life doesn't always fit into neat little boxes. As a result, the structured data always had to be supplemented by paper or microfilm storage. As technology performance has continued to improve, and prices have dropped, it was possible to bring into computing systems unstructured and semi-structured data.

1.3 Structured Data Technology Standards

SQL has been a standard of the American National Standards Institute since 1986. It is managed by InterNational Committee for Information Technology Standards (INCITS) Technical Committee DM 32 – Data Management and Interchange. The committee has two task groups, one for databases and the other for metadata. HP, CA, IBM, Microsoft, Oracle, Sybase (SAP) and Teradata all participate, as well as several federal government agencies. Both of the committee project documents have links to further information on each project. SQL became an International Organization for Standards (ISO) standard in 1987. The published standards are available for purchase from the ANSI eStandards Store, under the INCITS/ISO/IEC 9075 classification.

2. Semi-structured Data

Semi-structured data is a form of structured data that does not conform with the formal structure of data models associated with relational databases or other forms of data tables, but nonetheless, contains tags or other markers to separate semantic elements and enforce hierarchies of records and fields within the data. Therefore, it is also known as self-describing structure.

In semi-structured data, the entities belonging to the same class may have different attributes even though they are grouped together, and the attributes' order is not important.

Semi-structured data are increasingly occurring since the advent of the Internet where full-text documents and databases are not the only forms of data anymore, and different applications need a medium for exchanging information. In object-oriented databases, one often finds semi-structured data.

2.1 Types of Semi-structured data

2.1.1 XML

XML, other markup languages, email, and EDI are all forms of semi-structured data. OEM (Object Exchange Model) was created prior to XML as a means of self-describing a data structure. XML has been popularized by web services that are developed utilizing SOAP principles.

Some types of data described here as "semi-structured", especially XML, suffer from the impression that they are incapable of structural rigor at the same functional level as Relational Tables and Rows. Indeed, the view of XML as inherently semi-structured (previously, it was referred to as "unstructured") has handicapped its use for a widening range of data-centric applications. Even documents, normally thought of as the epitome of semi-structure, can be designed with virtually the same rigor as database schema, enforced by the XML schema and processed by both commercial and custom software programs without reducing their usability by human readers.

In view of this fact, XML might be referred to as having "flexible structure" capable of human-centric flow and hierarchy as well as highly rigorous element structure and data typing.

2.1.2 JSON

JSON or JavaScript Object Notation, is an open standard format that uses human-readable text to transmit data objects consisting of attribute–value pairs. It is used primarily to transmit data between a server and web application, as an alternative to XML. JSON has been popularized by web services developed utilizing REST principles.

There is a new breed of databases such as MongoDB and Couchbase that store data natively in JSON format, leveraging the pros of semi-structured data architecture.

2.2 Pros and Cons of Using a Semi-structured Data Format

2.2.1 Advantages

- Programmers persisting objects from their application to a database do not need to worry about object-relational impedance mismatch, but can often serialize objects via a light-weight library.
- Support for nested or hierarchical data often simplifies data models representing complex relationships between entities.
- Support for lists of objects simplifies data models by avoiding messy translations of lists into a relational data model.

2.2.2 Disadvantages

- The traditional relational data model has a popular and ready-made query language, SQL.
- Prone to "garbage in, garbage out"; by removing restraints from the data model, there is less fore-thought that is necessary to operate a data application.

3. Unstructured Data

Unstructured data (or unstructured information) refers to information that either does not have a pre-defined data model or is not organized in a pre-defined manner. Unstructured information is typically text-heavy, but may contain data such as dates, numbers, and facts as well. This results in irregularities and ambiguities that make it difficult to understand using traditional programs as compared to data stored in fielded form in databases or annotated (semantically tagged) in documents.

In 1998, Merrill Lynch cited a rule of thumb that somewhere around 80%-90% of all potentially usable business information may originate in unstructured form. This rule of thumb is not based on primary or any quantitative research, but nonetheless is accepted by some.

IDC and EMC project that data will grow to 40 zettabytes by 2020, resulting in a 50-fold growth from the beginning of 2010. The Computer World magazine states that unstructured information might account for more than 70%-80% of all data in organizations.

3.1 Background

The earliest research into business intelligence focused in on unstructured textual data, rather than numerical data. As early as 1958, computer science researchers like H.P. Luhn were particularly concerned with the extraction and classification of unstructured text. However, only since the turn of the century has the technology caught up with the research interest. In 2004, the SAS Institute developed the SAS Text Miner, which uses Singular Value Decomposition (SVD) to reduce a hyper-dimensional textual space into smaller dimensions for significantly more efficient machine-analysis. The mathematical and technological advances sparked by machine textual analysis prompted a number of business to research applications, leading to the development of fields like sentiment analysis, voice of the customer mining, and call center optimization. The emergence of Big Data in the late 2000s led to a heightened interest in the applications of unstructured data analytics in contemporary fields such as predictive analytics and root cause analysis.

3.2 Issues with terminology

The term is imprecise for several reasons:

(1) Structure, while not formally defined, can still be implied.

(2) Data with some form of structure may still be characterized as unstructured if its structure is not helpful for the processing task at hand.

(3) Unstructured information might have some structure (semi-structured) or even be highly structured but in ways that are unanticipated or unannounced.

3.3 Dealing with unstructured data

Techniques such as data mining, natural language processing (NLP), and text analytics provide different methods to find patterns in, or otherwise interpret, this information. Common techniques for structuring text usually involve manual tagging with metadata or part-of-speech tagging for further text mining-based structuring. The Unstructured Information Management Architecture (UIMA) standard provided a common framework for processing this information to extract meaning and create structured data about the information.

Software that creates machine-processable structure can utilize the linguistic, auditory, and visual structure that exist in all forms of human communication. Algorithms can infer this inherent structure from text, for instance, by examining word morphology, sentence syntax, and other small- and large-scale patterns. Unstructured information can then be enriched and tagged to address ambiguities and relevancy-based techniques then used to facilitate search and discovery. Examples of "unstructured data" may include books, journals, documents, metadata, health records, audio, video, analog data, images, files, and unstructured text such as the body of an e-mail message, Web page, or word-processor document. While the main content being conveyed does not have a defined structure, it generally comes packaged in objects (e.g. in files or documents) that themselves have structure and are thus a mix of structured and unstructured data, but collectively this is still referred to as "unstructured data". For example, an HTML web page is tagged, but HTML mark-up typically serves solely for rendering. It does not capture the meaning or function of tagged elements in ways that support automated processing of the information content of the page. XHTML tagging does allow machine processing of elements, although it typically does not capture or convey the semantic meaning of tagged terms.

Since unstructured data commonly occurs in electronic documents, the use of a content or document management system which can categorize entire documents is often preferred over data transfer and manipulation from within the documents. Document management thus provides the means to convey structure onto document collections.

Search engines have become popular tools for indexing and searching through such data, especially text.

📖 New Words

characteristic	[ˌkærɪktəˈrɪstɪk]	n.特性，特征
		adj.特有的，表示特性的，典型的
model	[ˈmɔdəl]	n.模型，原型
		vt.模仿
		v.模拟
memory	[ˈmeməri]	n.存储器，内存
commercially	[kəˈməːʃəli]	adv.商业上，贸易上
huge	[hjuːdʒ]	adj.巨大的，极大的，无限的
supplement	[ˈsʌplimənt]	n. & v.补充
microfilm	[ˈmaikrəufilm]	n.缩影胶片
		v.缩微拍摄
interchange	[ˌintəˈtʃeindʒ]	vt.交换
committee	[kəˈmiti]	n.委员会
participate	[pɑːˈtisipeit]	vi.参与，参加，分享，分担
conform	[kənˈfɔːm]	vt.使一致，使遵守，使顺从
		vi.符合
tag	[tæg]	n.标签，标记符
		vt.加标签于
marker	[ˈmɑːkə]	n.标记
separate	[ˈsepəreit]	adj.分开的，分离的；个别的，单独的
		v.分开，隔离，分散
hierarchy	[ˈhaiərɑːki]	n.层次，层级
entity	[ˈentiti]	n.实体
increasingly	[inˈkriːsiŋli]	adv.日益，愈加
medium	[ˈmiːdjəm]	n.媒体，媒介
		adj.中间的，中等的
popularize	[ˈpɔpjuləraiz]	v.普及
rigor	[ˈrigə]	n.严格，严密，精确
inherently	[inˈhiərəntli]	adv.天性地，固有地
epitome	[iˈpitəmi]	n.摘要
virtually	[ˈvəːtjuəli]	adv.事实上，实质上
schema	[ˈskiːmə]	n.模式，方案

flexible	['fleksəbl]	adj.灵活的，柔软的，能变形的
capable	['keipəbl]	adj.有能力的，能干的，有可能的
alternative	[ɔ:l'tə:nətiv]	n.二中择一，可供选择的办法、事物
		adj.选择性的，二中择一的
breed	[bri:d]	n.品种，种类
natively	['neitivli]	adv.本机地，本地地
mismatch	[mis'mætʃ]	vt.使配错，使配合不当
	['mismætʃ]	n.错配
nested	['nestid]	adj.嵌套的
simplify	['simplifai]	vt.单一化，简单化
restraint	[ris'treint]	n.抑制，制止，克制
irregularity	[i,regju'læriti]	n.不规则，无规律
ambiguity	[,æmbi'gju:iti]	n.含糊，不明确
potentially	[pə'tenʃəli]	adv.潜在地
background	['bækgraund]	n.背景，后台
hyperdimensional	[,haipə,dai'menʃənəl]	adj.多维的
spark	[spɑ:k]	v.发动，触发；激起，鼓舞
emergence	[i'mə:dʒəns]	n.浮现，露出，出现
terminology	[,tə:mi'nɔlədʒi]	n.术语学
imprecise	[,impri'sais]	adj.不严密的，不精确的
implied	[im'plaid]	ad.暗指的，含蓄的
unanticipated	[,ʌnæn'tisipeitid]	ad.不曾预料到的
auditory	[ɔ:ditəri]	ad.耳的，听觉的
visual	['vizjuəl]	ad.看的，视觉的，形象的
infer	[in'fə:]	v.推断
morphology	[mɔ:'fɔlədʒi]	n.词法，词态学
relevancy	['reləvənsi]	n.关联
capture	['kæptʃə]	n. & vt.捕获

Phrases

structured data	结构化数据
semi-structured data	半结构化数据
unstructured data	非结构化数据
fixed field	固定字段
data type	数据类型
data input	数据输入

fit into	适合
filing cabinet	档案柜
database management system	数据局管理系统
American National Standards Institute	美国国家标准协会
semantic element	语义元素
belong to	属于
full-text document	全文本文档
object-oriented database	面向对象数据库
markup language	标记语言
suffer from	忍受，遭受
data-centric application	以数据为中心的应用
garbage in, garbage out	垃圾进、垃圾出；无用数据入、无用数据出
rule of thumb	经验法则，大拇指规则
somewhere around	大约
result in	导致，产生
account for	占据
focused in on	着重于，关注
be concerned with	注重
predictive analytic	预测分析
root cause analysis	根源分析法
may be characterized as	可以称为
at hand	在手边，在附近，即将到来
deal with	处理，涉及，安排
part-of-speech tagging	词性标记，词类标识，词类标注
sentence syntax	句法，语句结构，句子结构
analog data	模拟数据
serve for	充当，用作
search engine	搜索引擎
search through …	把……仔细搜寻一遍

Abbreviations

INCITS (InterNational Committee for Information Technology Standards)	国际信息技术标准委员会
EDI (Electronic Data Interchange)	电子数据交换
OEM (Object Exchange Model)	对象交换模型
SOAP (Simple Object Access Protocol)	简单对象访问协议

JSON (Java Script Object Notation)	Java 脚本对象符号
REST (Representational State Transfer)	表述性状态传递
IDC（International Data Corporation）	国际数据公司
SVD (Singular Value Decomposition)	奇异值分解
NLP (Natural Language Processing)	自然语言处理
UIMA (Unstructured Information Management Architecture)	非结构化信息管理体系结构
HTML (HyperText Markup Language)	超文本标记语言
XHTML (Extensible HyperText Markup Language)	扩展超文本标记语言

Exercises

【Ex. 5】 根据课文内容回答以下问题。

1. What does structured data refer to?
2. What advantage does structured data have?
3. How is structured data often managed?
4. What is SQL? When did it become an International Organization for Standards (ISO) standard?
5. What is semi-structured data?
6. What are the types of semi-structured data mentioned in the text?
7. What are the disadvantages of using a semi-structured data format?
8. What does unstructured data refer to?
9. What techniques are used to deal with unstructured data?
10. Why is the use of a content or document management system which can categorize entire documents often preferred over data transfer and manipulation from within the documents?

参考译文

数 据 结 构

数据结构是组织和存储数据的特殊格式。一般数据结构类型包括数组、文件、记录、表、树等。所有数据结构的设计都是为了达到某一特定目的而组织数据，以便可以用适当的方式访问和工作。在计算机编程中，为了可以用多种算法工作，也可以选择或设计数据结构来存储数据。

1. 数组

（1）在数据存储中，数组是在多种设备上存储信息的方法。

（2）一般来说，数组是按照特定方法（例如，以列表或三维表）排列的许多项目。

（3）在计算机编程语言中，数组是具有相同属性、可以使用如加下标这样的技术分别访问的一组对象。

（4）在随机访问存储器中，数组是许多内存单元的排列。

2. 文件

（1）在数据处理中，文件是一些相关记录的集合。例如，可以把每个客户的记录放到一个文件中。依次地，每个记录由用于独立数据项的域组成，如客户姓名、客户编号、客户地址等。通过在每个记录相同域中提供同类信息（这样所有记录都一致），文件可方便地被计算机程序访问和处理。随着数据库的出现，使用这些术语已经不太重要了，而且它的重点在于用某一方法集合记录和域数据的表。在主机系统中，术语数据集通常与文件同义，但意味着它是可以由特定访问方式辨认的特定组织格式。取决于不同的操作系统，文件（和数据集）可以包含在一个类目、目录或文件夹中。

（2）在任一计算机系统，尤其是个人计算机中，文件是系统用户（包括系统自身及其应用程序）可用的数据实体，可以将其作为实体来处理（例如，从一个文件目录移动到另一个目录）。在自己的目录中，文件必须有唯一的名字。某些操作系统和应用程序通过给特定格式的文件特定的文件名后缀格式来描述文件。文件名后缀也称作文件扩展名。例如，程序或可执行文件有时给定或必须有".exe"后缀。一般情况下，后缀往往在操作系统允许的字符数以内尽可能地描述文件的格式。

3. 记录

（1）在计算机数据处理中，记录是排列的、以备程序处理的数据项的集合。多项记录可以组成文件或数据集。以记录形式组织的结构数据通常由定义记录的组织结构的编程语言规定，并/或由处理数据的应用程序来定义。通常，记录可以有固定的长度，或带有包含在记录内的长度信息的可变长度。

（2）在数据库中，记录——有时也称作行——是与特定实体相关的表中的一组域。例如，在一个客户联系信息表中，一行中通常包含这样的域：ID 号、姓名、街道地址、城市、电话号码等。

4. 表

在计算机编程语言中，表是用来组织信息的数据结构，就像在纸上一样。有多种不同的计算机关系表，用许多不同的方式来工作。下面列出比较普通的类型。

（1）在数据处理中，表也称作数组，是组织好的一组域。表可以存储相对不变的数据，也可被频繁更新。例如，包含在磁盘卷号中的表在写扇区时就被更新。

（2）在关系数据库中，表有时也称作文件，它把单一标题的信息组织到行和列中。例如，商业数据库通常包含客户信息表，该表中会用许多列来存储客户的账号、地址、电话号码等。数据的每个单一段（如账号）是表中的一个域。一列包含单个域中的全部项，如全部客户的电话号码。依次地，域被组织成记录，是信息的完整集合，如特定客户的信息集合，每条信息一行。这个规范化处理决定了如何有效地把数据组织到表中。

（3）决策表，通常称作真值表，可以用计算机或在纸上简单画出，它包含一系列的决策和做出决策所依据的标准。用于决策的各种可能出现的情况都必须列出，每种情况下采用的行动都应该被指定。一个简单的例子是：对于交通路口，通行决策也许可以表达为是与否，标准也许是红灯亮或绿灯亮。

决策表可以插入计算机程序中，以便根据不同的情况做出不同的决策。决策表的改变会反映在程序中。

（4）HTML 表用来在空间上组织网页元素，或建立可以按照表格形式更好地显示数据的数据结构，如列表或清单。

Unit 6

Text A

Basic Concepts of Database

1. Database

A database is a collection of information that is organized so that it can easily be accessed, managed and updated. Databases can be classified according to types of content: bibliographic, full-text, numeric and images.

In computing, databases are sometimes classified according to their organizational approach. The most prevalent approach is the relational database, a tabular database in which data is defined so that it can be reorganized and accessed in a number of different ways. A distributed database is one that can be dispersed or replicated among different points in a network. An object-oriented programming database is one that is congruent with the data defined in object classes and subclasses.

Computer databases typically contain aggregations of data records or files, such as sales transactions, product catalogs and inventories, and customer profiles. Typically, a database manager provides users the capabilities of controlling read/write access, specifying report generation and analyzing usage. Databases and database managers are prevalent in large mainframe systems, but are also present in smaller distributed workstation and mid-range systems such as the AS/400 and on personal computers.

2. Relational Database

A relational database is a collection of data items organized as a set of formally-described tables from which data can be accessed or reassembled in many different ways without having to reorganize the database tables. The relational database was invented by E. F. Codd at IBM in 1970.

The standard user and application program interface to a relational database is the structured query language (SQL). SQL statements are used both for interactive queries for information from a relational database and for gathering data for reports.

In addition to being relatively easy to create and access, a relational database has the important advantage of being easy to extend. After the original database creation, a new data category can be added without requiring that all existing applications be modified.

A relational database is a set of tables containing data fitted into predefined categories. Each table (which is sometimes called a relation) contains one or more data categories in columns. Each row contains a unique instance of data for the categories defined by the columns. For example, a typical business order entry database would include a table that described a customer with columns for name, address, phone number, and so forth. Another table would describe an order: product, customer, date, sales price, and so forth. A user of the database could obtain a view of the database that fitted the user's needs. For example, a branch office manager might like a view or report on all customers that had bought products after a certain date. A financial services manager in the same company could, from the same tables, obtain a report on accounts that needed to be paid.

When creating a relational database, you can define the domain of possible values in a data column and further constraints that may apply to that data value. For example, a domain of possible customers could allow up to ten possible customer names but be constrained in one table to allowing only three of these customer names to be specifiable.

The definition of a relational database results in a table of metadata or formal descriptions of the tables, columns, domains, and constraints.

3. SQL

SQL (Structured Query Language) is a standard language for making interactive queries from a database and updating a database such as IBM's DB2, Microsoft's Access and database products from Oracle, Sybase and Computer Associates. Although SQL is both an ANSI and an ISO standard, many database products support SQL with proprietary extensions to the standard language. Queries take the form of a command language that lets you select, insert, update, find out the location of data and so forth. There is also a programming interface.

4. Database Management System

A database management system (DBMS), sometimes just called a database manager, is a program that lets one or more computer users create and access data in a database. The DBMS manages user requests (and requests from other programs) so that users and other programs are free from having to understand where the data is physically located on storage media and, in a multi-user system, which else may also be accessing the data. In handling user requests, the DBMS ensures the integrity of the data (that is, making sure it continues to be accessible and is consistently organized as intended) and security (making sure only those with access privileges can access the data). The most typical DBMS is a relational database management system (RDBMS). A standard user and program interface is the Structured Query Language (SQL). A newer kind of DBMS is the object-oriented database management system (ODBMS).

A DBMS can be thought of as a file manager that manages data in databases rather than files in file systems. In IBM's mainframe operating systems, the non-relational data managers were (and are, because these legacy application systems are still used) known as access methods.

A DBMS is usually an inherent part of a database product. On PCs, Microsoft's Access is a popular example of a single- or small-group user DBMS. Microsoft's SQL Server is an example of a DBMS that serves database requests from multiple (client) users. Other popular DBMSs (these are all RDBMSs, by the way) are IBM's DB2, Oracle's line of database management products, and Sybase's products.

IBM's Information Management System (IMS) was one of the first DBMSs. A DBMS may be used by or combined with transaction managers, such as IBM's Customer Information Control System (CICS).

5. Distributed Database

A distributed database is a database in which portions of the database are stored on multiple computers within a network. Users have access to the portion of the database at their location so that they can access the data relevant to their tasks without interfering with the work of others.

6. DDBMS

A DDBMS (distributed database management system) is a centralized application that manages a distributed database as if it were all stored on the same computer. The DDBMS synchronizes all the data periodically, and in cases where multiple users must access the same data, ensures that updates and deletes performed on the data at one location will be automatically reflected in the data stored elsewhere.

7. Field

In a database table, a field is a data structure for a single piece of data. Fields are organized into records, which contain all the information within the table relevant to a specific entity. For example, in a table called customer contact information, telephone number would likely be a field in a row that would also contain other fields such as street address and city. The records make up the table rows and the fields make up the columns.

8. Record

In a database, a record (sometimes called a row) is a group of fields within a table that are relevant to a specific entity. For example, in a table called customer contact information, a row would likely contain fields such as: ID number, name, street address, city, telephone number and so on.

9. Table

In a relational database, a table (sometimes called a file) organizes the information about a single topic into rows and columns. For example, a database for a business would typically contain a table for customer information, which would store customers' account numbers, addresses, phone numbers, and so on as a series of columns. Each single piece of data (such as the account number) is a field in the table. A column consists of all the entries in a single field, such as the telephone numbers of all the customers. Fields, in turn, are organized as records, which are complete sets of information (such as the set of information about a particular customer), each of which comprises a row. The process of normalization determines how data will be most effectively organized into tables.

New Words

database	['deitəbeis]	n.数据库
organize	['ɔːgənaiz]	v.组织
classify	['klæsifai]	vt.分类，分等

bibliographer	[ˌbibliˈɔgrəfə]	adj.目录的
approach	[əˈprəutʃ]	n.方法，步骤，途径，通路
tabular	[ˈtæbjulə]	adj.制成表的，扁平的，表格式的，平坦的
		vi.列表，排成表格式
distributed	[disˈtribju:tid]	adj.分布式的
disperse	[disˈpə:s]	v.(使)分散，(使)散开
congruent	[ˈkɔngruənt]	adj.(与with连用)一致的，适合的
aggregation	[ægriˈgeiʃən]	n.集合，集合体，聚合
catalog	[ˈkætəlɔg]	n.目录，目录册
		v.编目录
capability	[ˌkeipəˈbiliti]	n.(实际)能力，性能，容量
analyze	[ˈænəlaiz]	vt.分析，分解
prevalent	[ˈprevələnt]	adj.普遍的，流行的
set	[set]	n.集合，集
reorganize	[ri:ˈɔ:gənaiz]	v.改组，再编制，改造
query	[ˈkwiəri]	v.询问，查询
extend	[iksˈtend]	v.扩充，延伸，伸展
instance	[ˈinstəns]	n.实例，例证
view	[vju:]	n.视图
domain	[dəuˈmein]	n.域，范围
constraint	[kənˈstreint]	n.约束，限制
specifiable	[ˈspesifaiəbl]	adj.能指定的；能详细说明的；能列举的
Oracle	[ˈɔrəkl]	n.美国甲骨文公司，主要生产数据库产品
command	[kəˈmɑ:nd]	n. & v.命令
insert	[inˈsə:t]	vt.插入
ensure	[inˈʃuə]	v.确保
privilege	[ˈprivilidʒ]	n.特权
inherent	[inˈhiərənt]	adj.固有的，内在的
client	[ˈklaiənt]	n.顾客，客户，委托人
centralize	[ˈsentrəlaiz]	vt.集聚，集中
synchronize	[ˈsiŋkrənaiz]	v.同步
periodically	[ˌpiəriˈɔdikəli]	adv.周期性地，定时性地
delete	[diˈli:t]	vt.删除
automatically	[ˌɔ:təˈmætikli]	adv.自动地
reflect	[riˈflekt]	v.反射，反映，表现
field	[fi:ld]	n.域
topic	[ˈtɔpik]	n.主题，题目

series	['siəri:z]	n.连续，系列
complete	[kəm'pli:t]	adj.完备的，完全的，完成的

Phrases

tabular database	表格数据库
distributed database	分布式数据库
customer profile	客户简介
find out	找出；发现

Abbreviations

SQL (structured query language)	结构化查询语言
IBM (International Business Machines Corporation)	国际商用机器公司
ANSI (American National Standards Institute)	美国国家标准协会
ISO (International Organization for Standardization)	国际标准化组织
DBMS (Database Management System)	数据库管理系统
RDBMS (Relational Database Management System)	关系型数据库管理系统
ODBMS (object-oriented database management system)	面向对象的数据库管理系统
IMS (Information Management System)	信息管理系统
CICS (Customer Information Control System)	客户信息管理系统
DDBMS (distributed database management system)	分布式数据库管理系统

Notes

[1] A relational database is a collection of data items organized as a set of formally-described tables from which data can be accessed or reassembled in many different ways without having to reorganize the database tables.

本句中，organized as a set of formally-described tables 是一个过去分词短语，作定语，修饰和限定 data items，它可以扩展成一个定语从句：which are organized as a set of formally-described tables。from which data can be accessed or reassembled in many different ways without having to reorganize the database tables 是一个介词前置的定语从句，修饰和限定 formally-described tables。

[2] A database management system (DBMS), sometimes just called a database manager, is a program that lets one or more computer users create and access data in a database.

本句中，sometimes just called a database manager 对 A database management system（DBMS）做进一步补充说明。that lets one or more computer users create and access data in a database 是一个定语从句，修饰和限定 a program。在该从句中，create and access data in a database 是一个不带 to 的动词不定式短语，作宾语 one or more computer users 的补足语。

英语中，在 make、let、have、see、hear、watch、notice、feel 等动词后面用动词不定式作宾语补足语时，不定式都不带 to。但当宾语补足语变成主语补足语时，to 不能省略。请看下例：

His boss often makes him work on weekends without extra pay.

他老板经常让他周末加班，却不给他额外报酬。

Let each man decide for himself.

让每个人自己决定。

Someone was heard to come up the stairs.

听见有人上楼。

[3] A DDBMS (distributed database management system) is a centralized application that manages a distributed database as if it were all stored on the same computer.

本句中，that manages a distributed database as if it were all stored on the same computer 是一个定语从句，修饰和限定 a centralized application。在该从句中，as if it were all stored on the same computer 是一个方式状语从句。

英语中，as if 和 as though 引导的方式状语从句一般要用虚拟语气。请看下例：

He talks as if he were a knowing-all.

他说起话来好像他是一个百事通。

[4] The DDBMS synchronizes all the data periodically, and in cases where multiple users must access the same data, ensures that updates and deletes performed on the data at one location will be automatically reflected in the data stored elsewhere.

本句中，in cases where multiple users must access the same data 作条件状语。that updates and deletes performed on the data at one location will be automatically reflected in the data stored elsewhere 是一个宾语从句，作 ensures 的宾语。在该从句中，performed on the data at one location 是一个过去分词短语，作定语，修饰和限定 updates and deletes。

[5] For example, in a table called customer contact information, telephone number would likely be a field in a row that would also contain other fields such as street address and city.

本句中，called customer contact information 是一个过去分词短语，作定语，修饰和限定 a table。that would also contain other fields such as street address and city 是一个定语从句，修饰和限定 a row。

Exercises

【Ex. 1】 根据课文内容判断以下叙述的正误。

1. A database is a collection of organized information.
2. A relational database is a tabular database in which data is defined so that it can be reorganized and accessed in a number of different ways.
3. A distributed database is one that can be dispersed or replicated at certain points in a network.
4. An object-oriented programming database is one that is congruent with the data defined in object classes and subclasses.
5. Databases and database managers are prevalent only in large mainframe systems.
6. The most typical DBMS is a distributed database management system.
7. A DBMS can be thought of as a file manager that manages data in databases.
8. The records make up the columns and the fields make up the table rows.

【Ex. 2】 根据课文内容填空。

1. According to types of content databases can be classified into _____, _____, _____ and images.
2. The relational database was invented by _____ in _____.
3. SQL stands for _____. It is a standard language for making interactive queries from a database and updating a database.
4. SQL statements are used both for _____ from a relational database and _____.
5. Queries take the form of a command language that lets you _____, _____, _____, find out the location of data and so forth. There is also _____.
6. A database management system (DBMS), sometimes just called _____, is a program that lets one or more computer users _____ in a database.
7. A standard user and program interface is _____. A newer kind of DBMS is _____.
8. A DDBMS (distributed database management system) is a centralized application that manages _____ as if it were all stored on the same computer.
9. In a database table, a field is _____. Fields are organized into _____.
10. In a database, a record is a group of fields within _____ that are relevant to _____.

【Ex. 3】 从题后的词组中选择与以下各条叙述意义最接近的词汇。

1. The type of computer processing where the user of the system communicates directly with the system to input data and instructions and receive output.
2. The boundary between two systems; a shared boundary between two systems.
3. The capability of having two or more jobs in the computer at the same time. Execution of the program is interleaved so that in a time interval each job will have been (partly) processed. Processing is not simultaneous.
4. A pictorial representation of processes and procedures for operation on data. A diagram that describes documents, procedures, processes, and equipment used in processing data in a specific application.
5. Performing tests and checks on input to ensure that the input operation is legal and that the input it self is correct. Pertaining to a wide variety of tests that can be applied to ensure the correctness of data being input to a computer system.

供选择的答案:

A. decision table
B. environment
C. flowchart
D. input/output system
E. input validation
F. integrated circuit
G. interactive computing
H. interface
I. Multiprogramming

【Ex. 4】 选择填空。

　　(1)　 analysis emphasizes the drawing of pictorial system models to document and validate both existing and/or proposed systems. Ultimately, the system models become the 　(2)　 for designing and constructing an improved system. 　(3)　 is such a technique. The emphasis in this technique is process-centered. Systems analysts draw a series of process models called 　(4)　. 　(5)　 is another such technique that integrates data and process concerns into constructs called objects.

供选择的答案:

1. A. Prototyping　　　B. Accelerated　　　C. Model-driven　　　D. Iterative
2. A. image　　　　　　B. picture　　　　　C. layout　　　　　　D. blueprint
3. A. Structured analysis　　　　　　　　　B. Information Engineering
   C. Discovery Prototyping　　　　　　　　D. Object-Oriented analysis
4. A. PERT　　　　　　B. DFD　　　　　　C. ERD　　　　　　　D. UML
5. A. Structured analysis　　　　　　　　　B. Information Engineering
   C. Discovery Prototyping　　　　　　　　D. Object-Oriented analysis

Text B

How Cloud Storage Works

Comedian George Carlin has a routine in which he talks about how humans seem to spend their lives accumulating "stuff." Once they've gathered enough stuff, they have to find places to store all of it. If Carlin were to update that routine today, he could make the same observation about computer information. It seems that everyone with a computer spends a lot of time acquiring data and then trying to find a way to store it.

For some computer owners, finding enough storage space to hold all the data they've acquired is a real challenge. Some people invest in larger hard drives. Others prefer external storage devices like thumb drives or compact discs. Desperate computer owners might delete entire folders worth of old files in order to make space for new information. But some are choosing to rely on a growing trend: cloud storage.

While cloud storage sounds like it has something to do with weather fronts and storm systems it really refers to saving data to an off-site storage system maintained by a third party. Instead of storing information to your computer's hard drive or other local storage device, you save it to a remote database. The Internet provides the connection between your computer and the database.

On the surface, cloud storage has several advantages over traditional data storage. For example, if you store your data on a cloud storage system, you'll be able to get to that data from any location that has Internet access. You wouldn't need to carry around a physical storage device or use the same computer to save and retrieve your information. With the right storage system, you could even allow other people to access the data, turning a personal project into a collaborative effort.

1. Cloud Storage Basics

There are hundreds of different cloud storage systems. Some have a very specific focus, such as storing Web e-mail messages or digital pictures. Others are available to store all forms of digital data. Some cloud storage systems are small operations, while others are so large that the physical equipment can fill up an entire warehouse. The facilities that house cloud storage systems are called data centers.

At its most basic level, a cloud storage system needs just one data server connected to the Internet. A client (e.g., a computer user subscribing to a cloud storage service) sends copies of files over the Internet to the data server, which then records the information. When the client wishes to retrieve the information, he or she accesses the data server through a Web-based interface. The server then either sends the files back to the client or allows the client to access and manipulate the files on the server itself.

Cloud storage systems generally rely on hundreds of data servers. Because computers occasionally require maintenance or repair, it's important to store the same information on multiple machines. This is called redundancy. Without redundancy, a cloud storage system couldn't ensure clients that they could access their information at any given time. Most systems store the same data on servers that use different power supplies. That way, clients can access their data even if one power supply fails.

Not all cloud storage clients are worried about running out of storage space. They use cloud storage as a way to create backups of data. If something happens to the client's computer system, the data survives off-site. It's a digital-age variation of "don't put all your eggs in one basket."

2. Examples of Cloud Storage

There are hundreds of cloud storage providers on the Web, and their numbers seem to increase every day. Not only are there a lot of companies competing to provide storage, but also the amount of storage each company offers to clients seems to grow regularly.

You're probably familiar with several providers of cloud storage services, though you might not think of them in that way. Here are a few well-known companies that offer some form of cloud storage:

- Google Docs allows users to upload documents, spreadsheets and presentations to Google's data servers. Users can edit files using a Google application. Users can also publish documents so that other people can read them or even make edits, which means Google Docs is also an example of cloud computing.
- Web e-mail providers like Gmail, Hotmail and Yahoo! Mail store e-mail messages on their own servers. Users can access their e-mail from computers and other devices connected to the Internet.
- Sites like Flickr and Picasa host millions of digital photographs. Their users create online photo albums by uploading pictures directly to the services' servers.
- YouTube hosts millions of user-uploaded video files.
- Web site hosting companies like StartLogic, Hostmonster and GoDaddy store the files and data for client Web sites.
- Social networking sites like Facebook and MySpace allow members to post pictures and other content. All of that content is stored on the respective site's servers.
- Services like Xdrive, MediaMax and Strongspace offer storage space for any kind of

digital data.

Some of the services listed above are free. Others charge a flat fee for a certain amount of storage, and still others have a sliding scale depending on what the client needs. In general, the price for online storage has fallen as more companies have entered the industry. Even many of the companies that charge for digital storage offer at least a certain amount for free.

3. Concerns about Cloud Storage

The two biggest concerns about cloud storage are reliability and security. Clients aren't likely to entrust their data to another company without a guarantee that they'll be able to access their information whenever they want and no one else will be able to get at it.

To secure data, most systems use a combination of techniques, including:

- Encryption, which means they use a complex algorithm to encode information. To decode the encrypted files, a user needs the encryption key. While it's possible to crack encrypted information, most hackers don't have access to the amount of computer processing power they would need to decrypt information.
- Authentication processes, which require to create a user name and password.
- Authorization practices—the client lists the people who are authorized to access information stored on the cloud system. Many corporations have multiple levels of authorization. For example, a front-line employee might have very limited access to data stored on a cloud system, while the head of human resources might have extensive access to files.

Even with these protective measures in place, many people worry that data saved on a remote storage system is vulnerable. There's always the possibility that a hacker will find an electronic back door and access data. Hackers could also attempt to steal the physical machines on which data are stored. A disgruntled employee could alter or destroy data using his or her authenticated user name and password. Cloud storage companies invest a lot of money in security measures in order to limit the possibility of data theft or corruption.

The other big concern, reliability, is just as important as security. An unstable cloud storage system is a liability. No one wants to save data to a failure-prone system, nor do they want to trust a company that isn't financially stable. While most cloud storage systems try to address this concern through redundancy techniques, there's still the possibility that an entire system could crash and leave clients with no way to access their saved data.

Cloud storage companies live and die by their reputations. It's in each company's best interests to provide the most secure and reliable service possible. If a company can't meet these basic client expectations, it doesn't have much of a chance—there are too many other options available on the market.

New Words

routine	[ruːˈtiːn]	n. 例行程序，常规
		adj. 常规的，例行的
accumulate	[əˈkjuːmjuleit]	v. 积聚，堆积
stuff	[stʌf]	n. 原料，材料
		vt. 塞满，填满，填充
observation	[ˌɔbzəːˈveiʃən]	n. 观察，观测；观察资料（或报告）
desperate	[ˈdespərit]	adj. 不顾一切的，拼死的，令人绝望的
folder	[ˈfəuldə]	n. 文件夹
e-mail	[ˈiːmeil]	n. 电子邮件
form	[fɔːm]	n. 形式，表格
		v. 形成，构成，排列，（使）组成
equipment	[iˈkwipmənt]	n. 装备，设备，器材，装置
house	[haus]	v. 给……提供地方；收藏；安置
survive	[səˈvaiv]	v. 幸免于，幸存，生还
variation	[ˌvɛəriˈeiʃən]	n. 变更，变化，变异，变种
regularly	[ˈregjuləli]	adv. 有规律地，有规则地；整齐地
upload	[ʌpˈləud]	vt. & n. 上传，上载
entrust	[inˈtrʌst]	v. 委托
crack	[kræk]	v. 攻破
hacker	[ˈhækə]	n. 黑客
decrypt	[diːˈkript]	v. 解密
password	[ˈpɑːswəːd]	n. 密码，口令
authorization	[ˌɔːθəraiˈzeiʃən]	n. 授权，认可
front-line	[frʌnt-lain]	adj. 前线的
extensive	[iksˈtensiv]	adj. 广大的，广阔的，广泛的
vulnerable	[ˈvʌlnərəbl]	adj. 易受攻击的
disgruntled	[disˈgrʌntld]	adj. 不满的，不高兴的
address	[əˈdres]	v. 解决，处理
crash	[kræʃ]	n. & v. 崩溃，垮台
reputation	[ˌrepju(ː)ˈteiʃən]	n. 名誉，名声

Phrases

cloud storage	云存储
storage space	存储空间

invest in	投资于
hard drive	硬盘驱动器
external storage device	外部存储设备
weather front	锋面，天气情况
storm system	风暴系统
thumb drive	U 盘
compact disc	光盘
have something to do with…	与……有点关系
offsite storage	异地存储，远程存储
third party	第三方
local storage device	本地存储设备
on the surface	表面上
physical storage device	物理存储设备
fill up	填补，装满
subscribing to	订购
power supply	电源
flat fee	固定费用
be worried about	为……忧虑，烦恼的
be familiar with	熟悉
social networking site	社交网络网站
sliding scale	浮动制计费，浮动费率制，按比例增减
multiple level	多层
protective measures	保全措施，保护措施
back door	后门

Exercises

【Ex. 5】 根据课文回答以下问题。

1. What is a real challenge for some computer owners?
2. What might desperate computer owners do in order to make space for new information?
3. What does cloud storage really refer to?
4. What does a cloud storage system need at its most basic level?
5. Why is it important to store the same information on multiple machines?
6. What do cloud storage clients use cloud storage as?
7. What does Google Docs allows users to do?
8. What do social networking sites like Facebook and MySpace allow members to do? Where is all that content stored?

9. What are the two biggest concerns mentioned in the passage about cloud storage?
10. What do most systems do to secure data?

参考译文

数据库基本概念

1. 数据库

数据库是信息的集合，这些信息被组织起来以便可以容易地访问、管理和更新。数据库可以按照其内容分为以下几类：书籍目录数据库、全文本数据库、数字数据库和图像数据库。

在计算领域中，数据库有时也按照其组织方法来分类。当前最流行的方法就是关系数据库，即一个定义数据的、以便可以用多种不同的方法来重新组织和访问的表格式数据库。分布式数据库是一个在网络中许多不同的地方分布或复制的数据库。面向对象编程数据库是一个适合用对象类和子类定义数据的数据库。

计算机数据库通常包含数据记录或文件的集合，如销售业务、产品目录和库存以及客户概况。通常，数据库管理程序给用户提供控制读/写访问、产生报表和分析使用情况的能力。数据库和数据库管理程序在大型机系统中非常普遍，但也出现在更小的分布式工作站和中等规模的系统中，如出现在 AS/400 或个人计算机中。

2. 关系数据库

关系数据库是数据项的集合，这些数据项组织为正式描述的表格的一个集合，其中的数据可以用多种方式访问或调整而无须重新组织数据库表。关系数据库由 E. F. Codd 于 1970 年在 IBM 创造。

关系数据库的标准用户和应用程序接口是结构化查询语言（SQL）。SQL 语句既可用于对关系数据库进行交互式信息查询，也可用于收集报表信息。

除了相对容易建立和访问之外，关系数据库的主要优点是容易扩展。建立了原始数据库后，可以增加新的数据库类别而无须对现有所有应用进行修改。

关系数据库是包含预设种类中数据的表格的集合。每个表（有时也叫作关系）按列包含一个或多个数据类。每行包括由列所定义的类型的唯一数据项。例如，一个典型的商务定单项数据库可以包括一个描述客户的表，该表列有客户姓名、地址、电话号码等。另一个表描述订单：产品、客户、日期、销售价格等。该数据库的用户可以获得他所需要的数据库概况。一个分部经理也许需要在某个日期之后购买产品的全部客户的概况或报表。同一公司的金融服务经理可以从同一表中获得需要支付的账号报表。

建立一个关系数据库后，可以在一个数据列中定义可能值的域以及未来可以应用到这些值的约束。例如，一个潜在客户域最多可以允许有 10 个客户的名称，但限制在一个表中只能列出 3 个这样的客户。

关系数据库的定义会产生一个元数据表或对该表、列、域和约束的正式描述。

3. SQL

SQL（结构化查询语言）是一个标准语言，用来对数据库进行交互式查询并更新数据库，如 IBM DB2、Microsoft Access 以及来自 Oracle、Sybase 的数据库产品和 Computer Associates。尽管 SQL 既是一个 ANSI 标准，也是一个 ISO 标准，但许多产品支持对标准语言的专门扩展的 SQL。请求的形式是命令行语言，可以让用户进行选择、插入、更新、找出数据的位置等。也有一个编程接口。

4. 数据库管理系统

数据库管理系统（DBMS）有时也叫作数据库管理器，是让一个或多个计算机用户建立和访问数据库中数据的程序。DBMS 管理用户查询（及来自其他程序的查询），这样用户和其他程序就不需要知道这些数据在介质中存储的物理位置，并且在多用户系统中，也不必知道还有谁可能正在访问这些数据。在处理用户查询时，DBMS 确保数据的完整性（也就是，确保可以持续地被访问并且一直按照预先要求组织好）和安全性（确保只有那些有访问权的用户才可以访问这些数据）。最典型的 DBMS 是关系数据库管理系统（RDBMS）。一个标准的用户和程序接口是结构化查询语言（SQL）。一个更新的 DBMS 是面向对象数据库管理系统（ODBMS）。

DBMS 可以被看作一个文件管理器，它管理数据库中的数据而不是文件系统中的文件。在 IBM 的大型机操作系统中，非关系数据管理器曾经（并且现在也是，因为这些老的应用系统仍然在使用）以访问方法而广为人知。

DBMS 通常是数据库产品的固有部分。在 PC 上，Microsoft Access 是单一或小组用户 DBMS 的一个流行范例。Microsoft SQL Server 是适用于多用户（客户）数据库查询的一个范例。其他流行的 DBMS（顺便说一下，这些全部都是 RDBMS）是 IBM 的 DB2、Oracle 的数据库管理产品线以及 Sybase 的产品。

IBM 的信息管理系统（IMS）是最初的 DBMS 之一。DBMS 也可被像 IBM 的客户信息管理系统（CICS）这样的业务管理程序使用，或与其结合使用。

5. 分布式数据库

分布式数据库是数据库中的某些部分存储在网络中的多个计算机中的数据库。用户可以在自己的位置访问该数据库的一部分，这样他们可以访问与其工作相关的数据而不会影响其他人的工作。

6. DDBMS

DDBMS（分布式数据库管理系统）是一个集中应用程序，管理一个分布式数据库，就像该数据库存储在同一计算机上一样。DDBMS 定期地保证所有数据的同步，并且在多个用户必须访问相同数据的情况下，确保在一个地方对数据的更新和删除在其他地方存储的数据中会自动反映出来。

7. 字段

在数据库表中，字段是用于单一数据块的数据结构。字段组成为记录，包括表中与特定实体相关的全部信息。例如，在一个叫作客户联系信息的表中，电话号码可能是一行中的一个字段，该行也包含了其他字段，如街道地址和城市。记录构成了表行而字段构成了列。

8. 记录

在数据库中，记录（有时也叫作行）是表中与一个特定实体相关的一组字段。例如，在一个叫作客户联系信息的表中，一行可能包括这样的字段：标识号、名字、街道地址、城市、电话号码等。

9. 表

在关系数据库中，表（有时叫作文件）把单一主题的信息组成为行和列。例如，一个商用数据库通常包括一个客户信息表，该表把客户账号、地址、电话号码等存储为一系列的列。每个单一的数据块（如账号）是表中的字段。一列由单一字段的全部实体组成，如全部客户的电话号码。字段依次被组织为记录，这就组成了信息的完整集合（如某一特定客户的信息集合），每个记录构成一行。这个规范处理过程决定了怎样将数据最有效地组织为表。

Unit 7

Text A

Data Warehouse Frequently Asked Questions

Q1. What is a data warehouse? How does it differ from a Database Management System (DBMS)?

A data warehouse is a database that provides users with data extracted from online transaction processing systems, batch systems, and externally syndicated data.

By contrast, a DBMS is software that controls the data in a database. It provides data security, data integrity, interactive queries, interactive data-entry and updating, and data independence.

Q2. How do I know if my organization needs a data warehouse?

Ideal candidates for a data warehouse display three common characteristics: They operate in a highly competitive industry; They have vast amounts of data; And they are struggling with the integration of widely dispersed data. If your organization fits this profile, it could well benefit from implementing a data warehouse.

Q3. What's involved in designing, building, and implementing a data warehouse?

Atre Group has identified 12 distinct steps in what we call our Data Warehouse Navigator. They are:

1. Determine users' needs
2. Determine DBMS server platform
3. Determine hardware platform
4. Information and data modeling
5. Construct metadata repository
6. Data acquisition and cleansing
7. Data transform, transport and populate
8. Determine middleware connectivity
9. Prototyping, querying and reporting
10. Data mining
11. Online analytic processing (OLAP)
12. Deployment and system management

Q 4. Which of the steps takes the longest?

It may vary from organization to organization, but in most of the situations, activities surrounding Extract, Transform and Load (ETL) are the most time consuming activities. There are multiple reasons for this:

1. Data is scattered in various disparate sources, and it is stored repeatedly in different files. Most organizations don't necessarily have decent documentation that is the authoritative source that says what is what. Identifying the correct sources of data to be used for a data warehouse is a daunting task.

2. In most organizations, there are various versions of a so called Meta Data repository. Meta Data is data about data. Sorting out the information in the Meta Data repositories is very time consuming.

3. Quality of data leaves a lot to be desired. Dirty data is either inaccurate data or inconsistent data. Once again, it is challenging to determine what is clean and what is dirty. In order to make this distinction, one needs to work together with business representatives who are knowledgeable in the business rules. The best business representatives are always busy. As a result, it is very difficult to get their attention.

4. It is time consuming to identify the data needed for analysis purposes to be used in a data warehouse.

5. The sub steps are: extraction, scrubbing, reconciling, aggregating, and summarizing the data. Each one of these sub steps is also time consuming.

And as a result, the entire process of ETL is the longest lasting step.

Q5. What is data mining? Is it a part of a data warehouse effort?

Data mining is finding patterns in the data that are not easily detectable by intuition or experience. Data mining could be a part of a data warehouse effort, or it could be a separate activity.

A major difference between a data warehouse and data mining is that most times, one uses summaries while using a data warehouse; whereas for data mining, detailed level data is needed. The patterns usually get lost when the data is summarized.

In a number of organizations, data mining is considered a part of the data warehouse effort.

Q6. What is OLAP?

OLAP stands for online Analytical Processing and is a technique for processing large amounts of data for the purposes of business analysis. The fundamental goal of OLAP is to exponentially improve the time it takes to query or read business data. It fundamentally differs from operational processing, commonly referred to as OLTP (On-Line Transaction Processing), which is built to achieve better write performance. OLAP Servers process data summaries to predetermine results of "What If" analysis. Normally, OLAP servers extract data from the data warehouse and then summarize and organize the data into multidimensional structures, commonly known as Cubes. The multidimensional data structures (or cubes) make it simple and efficient for users to formulate complex queries, arrange data on a report, switch from summary to detail data and filter or slice data into meaningful subsets.

Q7. What is warehouse appliance?

A data warehouse appliance is a combination of hardware and software product that is designed specifically for analytical processing. An appliance allows the purchaser to deploy a high-performance data warehouse right out of the box.

In a traditional data warehouse implementation, the database administrator (DBA) can spend a significant amount of time tuning and putting structures around the data to get the database to perform well. With a data warehouse appliance, however, it is the vendor who is responsible for simplifying the physical database design layer and making sure that the software is tuned for the hardware.

When a traditional data warehouse needs to be scaled out, the administrator needs to migrate all the data to a larger, more robust server. When a data warehouse appliance needs to be scaled out, the appliance can simply be expanded by purchasing additional pug-and-play components.

A data warehouse appliance comes with its own operating system, storage, database management system (DBMS) and software. It uses massively parallel processing (MPP) and distributes data across integrated disk storage, allowing independent processors to query data in parallel with little contention and redundant components to fail gracefully without harming the entire platform. Data warehouse appliances use Open Database Connectivity (ODBC), Java Database Connectivity (JDBC), and OLE DB interfaces to integrate with other extract-transform-load (ETL) tools and business intelligence (BI) or business analytic (BA) applications.

Currently, smaller data warehouse appliance vendors seem to be concentrating on adding functionality, such as in-memory analytics, to their products in order to compete with the mega-vendors. It is anticipated, however, that all appliance vendors will be impacted by the trend toward inexpensive, high-performance, scalable virtualized data warehouse implementations that use regular hardware and open source software.

New Words

online	[ˈɔnlain]	n.联机，在线式
syndicate	[ˈsindikit]	n.企业联合组织
candidate	[ˈkændidit]	n.候选人
vast	[vɑːst]	adj.巨大的，大量的
dispersed	[disˈpəːst]	adj.被分散的，散布的
construct	[kənˈstrʌkt]	vt.建造，构造，创立
populate	[ˈpɔpjuleit]	v.迁移
surrounding	[səˈraundiŋ]	n.围绕物，环境
		adj.周围的
scattered	[ˈskætəd]	adj.离散的，分散的
disparate	[ˈdispərit]	adj.全异的
repeatedly	[riˈpiːtidli]	adv.重复地，再三地
decent	[ˈdiːsnt]	adj.相当好的、像样的
authoritative	[ɔːˈθɔritətiv]	adj.权威的，命令的
daunting	[ˈdɔːntiŋ]	adj.使人畏缩的
inaccurate	[inˈækjurit]	adj.错误的，不准确的
representative	[ˌrepriˈzentətiv]	n.代表
scrub	[skrʌb]	v.净化，擦洗
reconcile	[ˈrekənsail]	vt.协调，理顺

pattern	['pætən]	n.式样，模式
detectable	[di'tektəbl]	adj.可发觉的，可看穿的
intuition	[,intju'iʃən]	n.直觉
exponential	[,ekspəu'nenʃəl]	n.指数
		adj.指数的，幂数的
predetermine	['pri:di'tə:min]	v.预定，预先确定
multidimensional	[,mʌltidi'menʃənl]	adj.多维的
cube	[kju:b]	n.立方体，立方
formulate	['fɔ:mjuleit]	vt.用公式表示，规则化
arrange	[ə'reindʒ]	v.安排，排列
filter	['filtə]	n.筛选
		vt.过滤
combination	[,kɔmbi'neiʃən]	n.结合，联合，合并
deploy	[di'plɔi]	v.展开，配置
significant	[sig'nifikənt]	adj.有意义的，重大的，重要的
migrate	[mai'greit]	vi.迁移，移动，移往
robust	[rə'bʌst]	adj.健壮的
harm	[hɑ:m]	vt.伤害，损害
		n.伤害，损害
functionality	[,fʌŋkəʃə'næliti]	n.功能性
anticipate	[æn'tisipeit]	vt.预期，期望
scalable	['skeiləbl]	adj.可升级的

Phrases

Frequently Asked Question (FAQ)	常见问题
differ from	不同
provide with …	给……提供，以……装备
data independence	数据独立性
competitive industry	竞争性产业
data modeling	数据建模
On line analytic processing (OLAP)	联机分析处理
Meta Data	元数据
dirty data	脏数据，废数据
business rule	商务规则
multidimensional data structure	多维数据结构
formulate complex queries	规则化复杂查询

right out of the box	开箱即用
be scaled out	出局
pug-and-play component	即插即用部件
massively parallel processing (MPP)	大规模并行处理
Open Database Connectivity (ODBC)	开放数据库互连
Java Database Connectivity (JDBC)	Java 数据库连接
business intelligence (BI)	商务智能，商业智能
business analytic (BA)	商业分析
concentrate on	集中，全神贯注于
in-memory analytic	存储器内分析
virtualized data warehouse	虚拟数据仓库

Abbreviations

OLTP (On Line Transaction Processing)	联机事务处理系统
OLE (Object Linking and Embedding)	对象连接与嵌入

Notes

[1] Most organizations don't necessarily have decent documentation that is the authoritative source that says what is what.

本句中，that is the authoritative source that says what is what 是一个定语从句，修饰和限定 decent documentation。在该从句中，that says what is what 也是一个定语从句，修饰和限定 the authoritative source。what is what 是宾语从句，作 says 的宾语。

[2] Identifying the correct sources of data to be used for a data warehouse is a daunting task.

本句中，Identifying the correct sources of data to be used for a data warehouse 是一个动名词短语，作主语。to be used for a data warehouse 是一个动词不定式短语，作定语，修饰和限定 the correct sources of data。

[3] In order to make this distinction, one needs to work together with business representatives who are knowledgeable in the business rules.

本句中，In order to make this distinction 是一个目的状语，修饰谓语 needs。who are knowledgeable in the business rules 是一个定语从句，修饰和限定 business representatives。

[4] A major difference between a data warehouse and data mining is that most times, one uses summaries while using a data warehouse; whereas for data mining, detailed level data is needed.

本句中，that most times, one uses summaries while using a data warehouse; whereas for data mining, detailed level data is needed 是一个表语从句。在该从句中，while using a data warehouse 做时间状语，修饰 uses；whereas 表示对比。

[5] It fundamentally differs from operational processing, commonly referred to as OLTP (On Line Transaction Processing), which is built to achieve better write performance.

本句中，It 指 OLAP。commonly referred to as OLTP (On Line Transaction Processing) 是一个过去分词短语，对 operational processing 进一步补充说明。which is built to achieve better write performance 是一个非限定性定语从句，对 OLTP 进行补充说明。

[6] With a data warehouse appliance, however, it is the vendor who is responsible for simplifying the physical database design layer and making sure that the software is tuned for the hardware.

本句中，it is the vendor who is responsible for simplifying the physical database design layer and making sure that the software is tuned for the hardware. 是 it 引导的强调句型。它强调的是主语 vendor。

英语中，It is/was +强调部分+that/who 从句也可以强调状语。例如：

It was last week that he bought that new computer.

It was in 1970 that E. F. Codd at IBM invented the relational database.

请注意：强调谓语时要用 do 或 did。例如：

Mike did send his manager an email yesterday, saying that he had fixed the printer.

Please do come earlier next time.

Exercises

【Ex.1】 根据课文内容回答问题。

1. What is a data warehouse?
2. What are the common characteristics ideal candidates for a data warehouse have?
3. What is the third step in what the so called data warehouse Navigator?
4. Which of the steps takes the longest in most of the situations?
5. What does one need to do in order to determine what data is clean and what is dirty?
6. What is data mining?
7. What is the major difference between a data warehouse and data mining?
8. What does OLPA stand for? What is it?
9. What is warehouse appliance?
10. With a data warehouse appliance, what is the vendor responsible for?

【Ex.2】 根据给出的汉语词义和规定的词类写出相应的英语单词。每词的首字母已给出。

vt.用公式表示，规则化	f___
adj.多余的，冗余的	r___
v.展开，配置	d___
adj.离散的，分散的	s___
n.批处理	b___
n.服务器	s___
vi.改变，转化，变换	t___
adj.不一致的，不协调的，矛盾的	i___
n.式样，模式	p___
n.联机，在线式	o___
n.知识库，仓库	r___
adj.全异的	d___
adj.错误的，不准确的	i___
adj.基础的，基本的	f___
adj.多维的	m___
n.指数	e___
adj.健壮的	r___
n.功能性	f___
adj.可升级的	s___
n.执行	i___

【Ex.3】 把下列句子翻译为中文。

1. You can chat to other people who are online.
2. Each summer a new batch of students tries to find work.
3. Are your products and services competitive? How about marketing?
4. My server is having problems this morning.
5. It is proverbially easier to destroy than to construct.
6. The photochemical reactions transform the light into electrical impulses.
7. Experimental results show algorithm is robust to resist normal and geometrical attack.
8. The feedback that comes from disparate industry and different area having different result.
9. The other fundamental consideration in the conception of a plan is function.
10. Populations tend to grow at an exponential rate.

【Ex.4】将下列词填入适当的位置（每词只用一次）。

| gather | compile | interactive | operations | product |
| rapidly | models | organization | figures | specified |

A decision support system (DSS) is a computer-based information system that supports business or organizational decision-making activities. DSSs serve the management, __(1)__, and planning levels of an __(2)__ and help to make decisions, which may be __(3)__ changing and not easily __(4)__ in advance.

DSSs include knowledge-based systems. A properly designed DSS is an __(5)__ software-based system intended to help decision makers __(6)__ useful information from a combination of raw data, documents, and personal knowledge, or business __(7)__ to identify and solve problems and make decisions.

Typical information that a decision support application might __(8)__ and present includes: inventories of information assets (including legacy and relational data sources, cubes, data warehouses, and data marts), comparative sales __(9)__ between one period and the next, projected revenue figures based on __(10)__ sales assumptions.

Text B

Data Backup

In information technology, a backup, or the process of backing up, refers to the copying and archiving of computer data so it may be used to restore the original after a data loss event. The verb form is to back up in two words, whereas the noun is backup.

Backups have two distinct purposes. The primary purpose is to recover data after its loss, be it by data deletion or corruption. The secondary purpose of backups is to recover data from an earlier time, according to a user-defined data retention policy, typically configured within a backup application for how long copies of data are required. Though backups represent a simple form of disaster recovery, and should be part of any disaster recovery plan, backups by themselves should not be considered a complete disaster recovery plan. One reason for this is that not all backup systems are able to reconstitute a computer system or other complex configuration such as a computer cluster, active directory server, or database server by simply restoring data from a backup.

Since a backup system contains at least one copy of all data considered worth saving, the data storage requirements can be significant. Organizing this storage space and managing the backup process can be a complicated undertaking. A data repository model may be used to provide structure to the storage. Nowadays, there are many different types of data storage devices that are useful for making backups. There are also many different ways in which these devices can be arranged to provide geographic redundancy, data security, and portability.

Before data are sent to their storage locations, they are selected, extracted, and manipulated. Many different techniques have been developed to optimize the backup procedure. These include optimization for dealing with open files and live data sources as well as compression, encryption, and deduplication, among others. Every backup scheme should include dry runs that validate the reliability of the data being backed up. It is important to recognize the limitations and human factors involved in any backup scheme.

1. Selection and extraction of data

A successful backup job starts with selecting and extracting coherent units of data. Most data on modern computer systems is stored in discrete units, known as files. These files are organized into file systems. Files that are actively being updated can be thought of as "live" and present a challenge to back up. It is also useful to save metadata that describes the computer or the file system being backed up.

Deciding what to back up at any given time is a harder process than it seems. By backing up too much redundant data, the data repository will fill up too quickly. Backing up an insufficient amount of data can eventually lead to the loss of critical information.

1.1 Files

1.1.1 Copying files

With file-level approach, making copies of files is the simplest and most common way to perform a backup. A means to perform this basic function is included in all backup software and all operating systems.

1.1.2 Partial file copying

Instead of copying whole files, one can limit the backup to only the blocks or bytes within a file that have changed in a given period of time. This technique can use substantially less storage space on the backup medium, but requires a high level of sophistication to reconstruct files in a restore situation. Some implementations require integration with the source file system.

1.1.3 Deleted files

To prevent the unintentional restoration of files that have been intentionally deleted, a record of the deletion must be kept.

1.2 File systems

1.2.1 File system dump

Instead of copying files within a file system, a copy of the whole file system itself in block-level can be made. This is also known as a raw partition backup and is related to disk imaging. The process usually involves unmounting the file system and running a program like dd (Unix). Because the disk is read sequentially and with large buffers, this type of backup can be much faster than reading every file normally, especially when the file system contains many small files, is highly fragmented, or is nearly full. But because this method also reads the free disk blocks that contain no useful data, this method can also be slower than conventional reading, especially when the file system is nearly empty. Some file systems provide a "dump" utility that reads the disk sequentially for high performance while skipping unused sections. The corresponding restore utility can selectively restore individual files or the entire volume at the operator's choice.

1.2.2 Identification of changes

Some file systems have an archive bit for each file that says it was recently changed. Some backup software looks at the date of the file and compares it with the last backup to determine whether the file was changed.

1.2.3 Versioning file system

A versioning file system keeps track of all changes to a file and makes those changes accessible to the user. Generally this gives access to any previous version, all the way back to the file's creation time. An example of this is the Wayback versioning file system for Linux.

1.3 Live data

If a computer system is in use while it is being backed up, the possibility of files being open for reading or writing is real. If a file is open, the contents on disk may not correctly represent what the owner of the file intends. This is especially true for database files of all kinds. The term fuzzy backup can be used to describe a backup of live data that looks like it ran correctly, but does not represent the state of the data at any single point in time. This is because the data being backed up changed in the period of time between when the backup started and when it finished. For databases in particular, fuzzy backups are worthless.

1.3.1 Snapshot backup

A snapshot is an instantaneous function of some storage systems that presents a copy of the file system as if it were frozen at a specific point in time, often by a copy-on-write mechanism. An effective way to back up live data is to temporarily quiesce them (e.g. close all files), take a snapshot, and then resume live operations. At this point the snapshot can be backed up through normal methods. While a snapshot is very handy for viewing a filesystem as it was at a different point in time, it is hardly an effective backup mechanism by itself.

1.3.2 Open file backup

Many backup software packages feature the ability to handle open files in backup operations. Some simply check for openness and try again later. File locking is useful for regulating access to open files.

When attempting to understand the logistics of backing up open files, one must consider that the backup process could take several minutes to back up a large file such as a database. In order to back up a file that is in use, it is vital that the entire backup represent a single-moment snapshot of the file, rather than a simple copy of a read-through.

1.3.3 Cold database backup

During a cold backup, the database is closed or locked and not available to users. The data files do not change during the backup process so the database is in a consistent state when it is returned to normal operation.

1.3.4 Hot database backup

Some database management systems offer a means to generate a backup image of the database while it is online and usable (hot). This usually includes an inconsistent image of the data files plus a log of changes made while the procedure is running. Upon a restore, the changes in the log files are reapplied to bring the copy of the database up-to-date (the point in time at which the initial hot backup ended).

2. Managing the backup process

As long as new data are being created and changes are being made, backups will need to be performed at frequent intervals. Individuals and organizations with anything from one computer to thousands of computer systems all require protection of data. The scales may be very different, but the objectives and limitations are essentially the same. Those who perform backups need to know how successful the backups are, regardless of scale.

2.1 Objectives

2.1.1 Recovery point objective (RPO)

The point in time that the restarted infrastructure will reflect. Essentially, this is the roll-back that will be experienced as a result of the recovery. The most desirable RPO would be the point just prior to the data loss event. Making a more recent recovery point achievable requires increasing the frequency of synchronization between the source data and the backup repository.

2.1.2 Recovery time objective (RTO)

The amount of time elapsed between disaster and restoration of business functions.

2.2 Implementation

2.2.1 Scheduling

Using a job scheduler can greatly improve the reliability and consistency of backups by removing part of the human element. Many backup software packages include this functionality.

2.2.2 Authentication

Over the course of regular operations, the user accounts and/or system agents that perform the backups need to be authenticated at some level. The power to copy all data off of or onto a system requires unrestricted access. Using an authentication mechanism is a good way to prevent the backup scheme from being used for unauthorized activity.

2.2.3 Chain of trust

Removable storage media are physical items and must only be handled by trusted individuals. Establishing a chain of trusted individuals (and vendors) is critical to defining the security of the data.

New Words

archive	['ɑ:kaiv]	vt. 存档
		n. 档案文件
recover	[ri'kʌvə]	vt. 重新获得，恢复
disaster	[di'zɑ:stə]	n. 灾难，天灾，灾祸
reconstitute	[,ri:'kɔnstitju:t]	vt. 重新组成，重新设立
redundancy	[ri'dʌndənsi]	n. 冗余
portability	[,pɔ:tə'biləti]	n. 可携带，轻便
compression	[kəm'preʃən]	n. 浓缩，压缩
deduplication	[di,dju:pli'keiʃən]	n. 数据去重，删除重复数据
extraction	[iks'trækʃən]	n. 抽出，取出
discrete	[dis'kri:t]	adj. 不连续的，离散的
file	[fail]	n. 文件
metadata	['metədeitə]	n. 元数据
redundant	[ri'dʌndənt]	adj. 多余的
insufficient	[,insə'fiʃənt]	adj. 不足的，不够的
		n. 不足
reconstruct	[,ri:kən'strʌkt]	v. 重建
unintentional	[ʌnin'tenʃənl]	n. 不是故意的，无心的
buffer	['bʌfə]	n. 缓冲区，缓冲器

fragmented	[fræg'mentid]	*adj.* 成碎片的，片断的
fuzzy	['fʌzi]	*adj.* 模糊的
worthless	['wə:θlis]	*adj.* 无价值的，无益的
snapshot	['snæpʃɔt]	*n.* 快照
instantaneous	[,instən'teinjəs]	*adj.* 瞬间的，即刻的，即时的
inconsistent	[,inkən'sistənt]	*adj.* 不一致的，不协调的，矛盾的
up-to-date	['ʌptə'deit]	*adj.* 直到现在的，最近的，当代的
unrestricted	[,ʌnris'triktid]	*adj.* 无限制的，自由的

Phrases

data retention	数据保持
data repository model	数据仓库模型
dry run	演习，排练
disk imaging	磁盘镜像
copy-on-write	写时拷贝，写时复制
software package	软件包，程序包
file locking	文件锁定
regardless of	不管，不顾
job scheduler	作业安排

Abbreviations

| RPO (Recovery Point Objective) | 恢复点目标 |
| RTO (Recovery Time Objective) | 复原时间目标 |

Exercises

【Ex.5】 根据课文内容回答问题。

1. What does a backup refer to in information technology?
2. How many distinct purposes do backups have? What are they?
3. What is one reason that backups by themselves should not be considered a complete disaster recovery plan?
4. What does a successful backup job start with?
5. What is the simplest and most common way to perform a backup with file-level approach?
6. What is also known as a raw partition backup?

7. What does a versioning file system do?
8. What can the term fuzzy backup be used to do?
9. What is vital in order to back up a file that is in use?
10. What is a good way to prevent the backup scheme from being used for unauthorized activity?

参考译文

数据仓库常见问题

Q1. 什么是数据仓库？它与数据库管理系统（DBMS）有何不同？

数据仓库是一个数据库，为用户提供从联机事务处理系统、批处理系统以及外部联合数据中提取的数据。

相比之下，数据库管理系统是控制数据库中数据的软件。它提供数据安全性、数据完整性、交互式查询、交互式的数据录入和更新，以及数据的独立性。

Q2. 怎么知道我的组织是否需要一个数据仓库？

使用数据仓库的理想人选显示三个共同的特点：他们处在一个高度竞争的行业、他们要处理大量的数据以及他们正努力对广泛分散的数据进行整合。如果您所在的组织符合这些特征，就可以从实施数据仓库中获益。

Q3. 如何设计、构建和实施数据仓库？

Atre集团已确定了12个明确的步骤，称为"数据仓库导航"。它们是：
（1）确定用户的需求
（2）确定DBMS服务器平台
（3）确定硬件平台
（4）信息和数据建模
（5）构建元数据存储库
（6）数据采集和清理
（7）数据变换、传输和迁移
（8）确定中间件连接
（9）原型、查询和报告

（10）数据挖掘
（11）在线分析处理（OLAP）
（12）部署和系统管理

Q4. 哪一个步骤花费的时间最长？

可能因组织而异，但在大多数的情况下，提取、转换和加载（ETL）最耗费时间。这有多种原因：

（1）数据分散在各种不同数据源并且重复存储在不同的文件中。大多数组织无需权威的文件来确定数据的准确来源。识别用于数据仓库的正确数据源是一项艰巨的任务。

（2）在大多数组织中，有各种版本的所谓的元数据存储库。元数据是关于数据的数据。整理元数据库中的信息非常耗时。

（3）数据的质量有很多不足之处。不准确的数据或不一致的数据都是脏数据。另外，确定哪些数据是干净的、哪些是脏的颇具挑战性。为了找出区别，需要与熟悉业务规则的业务代表共同努力。最好的业务代表总是很忙。因而，要获得他们的关照非常困难。

（4）识别数据仓库中需要分析的数据很费时。

（5）这一子步骤是：提取、整理、协调、集中和汇总数据。其中每个步骤都需要时间。

因此，ETL 的整个过程持续时间最长。

Q5. 什么是数据挖掘？它是数据仓库工作的一部分吗？

数据挖掘是在不易凭直觉或经验察觉的数据中寻找模式。数据挖掘可以是数据仓库工作的一部分，也可以是一个单独的活动。

数据仓库和数据挖掘的主要区别是：在大多数情况下，使用数据仓库时使用概要数据，而数据挖掘需要详细数据。使用概要性的数据通常会丢失模式。

在若干组织中，认为数据挖掘是一个数据仓库工作的一部分。

Q6. OLAP是什么？

OLAP 代表联机分析处理，它是处理大量数据的一种技术，目的是为了进行业务分析。OLAP 的基本目标是指数级地减少查询或阅读业务数据所花费的时间。它与通常被称为 OLTP（联机事务处理）的操作处理根本不同，OLTP 的建立是为了实现更好的写入性能。OLAP 服务器处理数据摘要以便预先确定"要是……又怎样"的分析结果。通常情况下，OLAP 服务器从数据仓库中提取数据，然后汇总数据并将其组织成多维结构，俗称立方体。多维数据结构（或立方体）简单、高效地实现用户的规则化复杂查询、排列报表数据、把概要数据转换为详细数据并将数据过滤或划分为有意义的子集。

Q7. 仓库设备是什么？

数据仓库设备结合硬件和软件产品，专门用于分析处理。一个设备允许买方部署一个高性能的数据仓库，实现开箱即用。

在传统的数据仓库实施中，数据库管理员（DBA）要花费大量的时间调整和设置来自数据库的数据结构，使其性能良好。但是，使用数据仓库设备，供应商负责简化物理数据库设计层，并确保软件与硬件相适应。

当传统的数据仓库需要扩展时，管理员需要把所有数据迁移到一个更大的、更健壮的服务器上。当需要扩展数据仓库设备时，该设备可以简单地通过购买额外的即插即用组件来实现扩展。

数据仓库设备都带有自己的操作系统、存储器、数据库管理系统（DBMS）和软件。它采用大规模并行处理（MPP）和将数据分布到集成磁盘存储器，这样独立的处理器就可以并行查询数据，很少有争用和冗余组件失效，不会损害整个平台。数据仓库设备使用开放式数据库连接（ODBC）、Java 数据库连接（JDBC）和 OLE DB 接口并与提取-转换-加载（ETL）工具和商业智能（BI）或商业分析（BA）应用相集成。

目前，较小的数据仓库设备厂商似乎注重增加产品功能（如内存分析），以便能与大型供应商竞争。然而，据预计，所有的厂商都朝着以下趋势发展：价格低廉、高性能、使用普通硬件和开源软件的可扩展虚拟化数据仓库。

Unit 8

Text A

Data Preprocessing

Data preprocessing is a data mining technique that involves transforming raw data into an understandable format. Real-world data is often incomplete, inconsistent, and/or lacking in certain behaviors or trends, and is likely to contain many errors. Data preprocessing is a proven method of resolving such issues. Data preprocessing prepares raw data for further processing. Data goes through a series of steps during preprocessing.

1. Data Cleansing

Data cleansing, also known as data scrubbing, is the process of ensuring that a set of data is correct and accurate. During this process, records are checked for accuracy and consistency, and they are either corrected or deleted as necessary. This can occur within a single set of records or between multiple sets of data that need to be merged or that will work together.

At its most simple form, data cleansing involves a person or persons reading through a set of records and verifying their accuracy. Typos and spelling errors are corrected, mislabeled data is properly labeled and filed, and incomplete or missing entries are completed. These operations often purge out-of-date or unrecoverable records so that they do not take up space and cause inefficient operations.

In more complex operations, data cleansing can be performed by computer programs. These programs can check the data with a variety of rules and procedures decided upon by the user. A program could be set to delete all records that have not been updated within the previous five years, correct any misspelled words and delete any duplicate copies. A more complex program might be able to fill in a missing city based on a correct postal code or change the prices of all items in a database to another type of currency.

2. Data Integration

Data integration is the merging of multiple data sources into a single data source. This practice is often very time-consuming and involved, as the different data sources are likely incompatible with one another. Things as simple as different column names on a spreadsheet are enough to require date reformatting. This process is most common in situations where two groups started with no connection, but are placed together after they have worked independently. Data integration has become a more important topic due to the prevalence of free data sources and online databases.

The data part of data integration can be almost anything as long as it is stored in a computer system. The actual content of the data is rarely as important as the way in which the data is stored. Most of the time, the data is kept in databases, organized systems of information. These systems contain unique entries and fields that allow users to find information quickly.

The biggest hurdle to any data integration process is the data itself. In many cases, when the data was first set up, there was no intention of ever merging the dataset with another. This means that even though two datasets may refer to the same thing, they are totally incompatible.

Nearly anything will make databases incompatible. Something as simple as a difference in presentation, such as field order or column width, can be enough to prevent an easy merger. When the data is significantly different, such as one database that contains more or less information, the merging is much more difficult.

The two situations that call for data integration more than any other are in the business and the research fields. In the business world, merging departments or companies requires combining the previously separate information into a single structure. This form of integration is generally very difficult unless the original groups used similar software and had similar information goals.

When data integration is performed for research purposes, it generally goes much smoother. When one researcher gives access to his information to another, the two parties are generally looking into the same process. This means they will use similar methods to catalog and store their data.

In the past, data integration was a relatively minor area of data studies, but this has changed since the early part of the 21st century. With free online databases becoming more popular and accurate, companies are scrambling to get their information in a sharable format. This allows them to both release their information in a public form and to integrate private versions of well-known public interfaces into their systems.

3. Data Transformation

Data transformation is the process of converting information or data from one format to another format. While the strategy is often thought of in terms of converting documents from one format to another, data transformations may also involve converting programs from one type of computer language to a different format in order to allow the program to run on a specific platform. The actual transformation may involve converting multiple data streams into a common format, or converting a single format into multiple different forms for use across a wide spectrum of platforms.

The process of data transformation involves the use of what is known as SQL, or structured query language. SQL is the computer language that is responsible for managing the information that resides in some type of data management system.

In actual use, data transformation involves the use of an executable program that is capable of reading the base or original language of the data, and identifying the language or languages that the data must translate into in order to be used by other programs. Once the mapping for the transformation is accomplished, the program then converts the data into the single or multiple formats desired, and distributes the converted data accordingly. With many applications, this takes place in a matter of seconds.

A similar process is known as data mediation. Like data transformation, the idea is to make data in one format to be usable in another format. One difference with mediation is that the data mapping process involves the creation of what is known as a data model, serving as an intermediary between the two formats involved, rather than the direct translation that occurs with the transformation of information.

As with many types of computer technology, data transformation is a process that is continually evolving as new programs help to increase the efficiency and scope of how information can be translated. As more programs and formats are included in this process, the ability to share data across many different platforms that were once totally incompatible has increased significantly. In a global setting where collaborators may not always make use of the same programs or languages as the foundation for their data systems, these continual improvements mean that there is significantly less time needed to manually translate and enter data between systems.

4. Data Reduction

Data reduction is the transformation of numerical or alphabetical digital information derived empirically or experimentally into a corrected, ordered, and simplified form. The basic concept is the reduction of multitudinous amounts of data down to the meaningful parts.

When information is derived from instrument readings there may also be a transformation from analog to digital form. When the data are already in digital form the "reduction" of the data typically involves some editing, scaling, coding, sorting, collating, and producing tabular summaries. When the observations are discrete but the underlying phenomenon is continuous, smoothing and interpolation are often needed. Often the data reduction is undertaken in the presence of reading or measurement errors. Some idea of the nature of these errors is needed before the most likely value may be determined.

An example in astronomy is the data reduction in the Kepler satellite. This satellite records 95-megapixel images once every six seconds, generating tens of megabytes of data per second, which is orders of magnitudes more than the downlink bandwidth of 550 KBps. The on-board data reduction encompasses co-adding the raw frames for thirty minutes, reducing the bandwidth by a factor of 300. Furthermore, interesting targets are preselected and only the relevant pixels are processed, which is 6% of the total. This reduced data is then sent to Earth where it is processed further.

Research has also been carried out on the use of data reduction in wearable (wireless) devices for health monitoring and diagnosis applications. For example, in the context of epilepsy diagnosis, data reduction has been used to increase the battery lifetime of a wearable EEG device by selecting, and only transmitting, EEG data that is relevant for diagnosis and discarding background activity.

New Words

preprocess	[priː'prəuses]	vt.预加工，预处理
transform	[træns'fɔːm]	vt.转换，改变；使……变形
		vi.改变，转化，变换
understandable	[ʌndə'stændəbl]	adj.可以理解的，能懂的
incomplete	[,inkəm'pliːt]	adj.不完的，不完善的
accurate	['ækjurit]	adj.正确的，精确的
accuracy	['ækjurəsi]	n.精确性，正确度
consistency	[kən'sistənsi]	n.一致性
verify	['verifai]	vt.检验，校验
typos	['taipəus]	n.打字稿

mislabel	[mis'leibl]	v.贴错标签
purge	[pə:dʒ]	n.净化，清除 v.(使)净化，清除
out-of-date	['autəv'deit]	adj.过期的，过时的，落伍的
unrecoverable	[ʌnri'kʌvərəbl]	adj.不可恢复的
inefficient	[ˌini'fiʃənt]	adj.效率低的，效率差的
update	[ʌp'deit]	v.更新，校正
	[ʌp'deit]	n.更新；现代化；更新的信息
misspell	[mis'spel]	vt.拼错
integration	[ˌinti'greiʃən]	n.整合，综合
time-consuming	['taimkən,sju:miŋ]	adj.耗时的
involved	[in'vɔlvd]	adj.棘手的；有关的
incompatible	[ˌinkəm'pætəbl]	adj.不兼容的，矛盾的，不调和的
reformat	[ri'fɔ:mæt]	vt.重定格式，重新格式化
connection	[kə'nekʃən]	n.连接，关系
independently	[indi'pendəntli]	adv.独立地
prevalence	['prevələns]	n.流行
unique	[ju'ni:k]	adj.唯一的，独特的
hurdle	['hə:dl]	n.障碍
sharable	['ʃɛərəbl]	adj.可共享的，可分享的，可分担的
transformation	[ˌtrænsfə'meiʃən]	n.变化，转化
convert	[kən'və:t]	vt.使转变，转换
strategy	['strætidʒi]	n.策略
spectrum	['spektrəm]	n.频谱，波谱；范围
executable	['eksikju:təbl]	adj.可执行的，可实行的
mapping	['mæpiŋ]	n.映射
accomplish	[ə'kɔmpliʃ]	vt.完成，达到，实现
distribute	[dis'tribju(:)t]	vt.分发，分配，分布，分类，分区
continually	[kən'tinjuəli]	adv.不断地，频繁地
collaborator	[kə'læbəreitə]	n.合作者
alphabetical	[ˌælfə'betikəl]	adj.字母的
experimentally	[iksperi'mentəli]	adv.实验上，用实验方法
simplify	['simplifai]	vt.单一化，简单化
instrument	['instrəmənt]	n.工具，手段，器具
collate	[kɔ'leit]	v.比较
observation	[ˌɔbzə'veiʃən]	n.观察，观测；观察资料（或报告）
phenomenon	[fi'nɔminən]	n.现象

smooth	[smuːð]	*adj.*平滑的，平稳的，流畅的
		*vt.*使光滑
		*vi.*变平滑
interpolation	[inˌtəːpəˈleiʃən]	*n.*插补
astronomy	[əˈstrɔnəmi]	*n.*天文学
satellite	[ˈsætəlait]	*n.*人造卫星
megapixel	[ˈmegəpiksəl]	*n.*兆像素
megabyte	[ˈmegəbait]	*n.*兆字节
magnitude	[ˈmægnitjuːd]	*n.*大小，数量，量级
downlink	[ˈdaunliŋk]	*n.*下行链路
encompass	[inˈkʌmpəs]	*v.*包含，包括
factor	[ˈfæktə]	*n.*因素，要素
target	[ˈtɑːgit]	*n.*目标
preselect	[ˈpriːsiˈlekt]	*vt.*预先选择
pixel	[ˈpiksəl]	*n.*像素
wireless	[ˈwaiəlis]	*adj.*无线的
diagnosis	[ˌdaiəgˈnəusis]	*n.*诊断
discard	[disˈkɑːd]	*vt.*丢弃，抛弃，放弃

Phrases

data preprocessing	数据预处理
data mining	数据挖掘
raw data	原始数据
lacking in	缺少
data cleansing	数据清理
spelling error	拼写错误
take up	占据
duplicate copy	副本，复制本
postal code	邮政编码
data integration	数据整合
as long as	只要，在……的时候
set up	建立
look into	浏览，观察
data transformation	数据转换
data stream	数据流
be responsible for …	为……负责，形成……的原因

data management system	数据管理系统
be capable of	能够
translate into	转化为，翻译成
a matter of	大约，大概
data mediation	数据调节
data reduction	数据缩减
be derived from	源自于

✎ Abbreviations

KBps (Kilo-Bytes Per Second)	每秒千字节
EEG (Electro EncephaloGram)	脑电图

✎ Notes

[1] A program could be set to delete all records that have not been updated within the previous five years, correct any misspelled words and delete any duplicate copies.

本句中，to delete all records that have not been updated within the previous five years, correct any misspelled words and delete any duplicate copies 是动词不定式短语，作目的状语。that have not been updated within the previous five years 是一个定语从句，修饰和限定 all records。

[2] This process is most common in situations where two groups started with no connection, but are placed together after they have worked independently.

本句中，where two groups started with no connection, but are placed together after they have worked independently 是一个定语从句，修饰和限定 situations。

[3] This means that even though two datasets may refer to the same thing, they are totally incompatible.

本句中，that even though two datasets may refer to the same thing, they are totally incompatible 是一个宾语从句。在该从句中，even though two datasets may refer to the same thing 是一个让步状语从句，修饰谓语 are totally incompatible。even though 的意思是"即使""尽管"。

[4] While the strategy is often thought of in terms of converting documents from one format to another, data transformations may also involve converting programs from one type of computer language to a different format in order to allow the program to run on a specific platform.

本句中，While the strategy is often thought of in terms of converting documents from one format to another 是一个让步状语从句。in order to allow the program to run on a specific platform 作目的状语。

[5] In actual use, data transformation involves the use of an executable program that is capable of reading the base or original language of the data, and identifying the language or languages that the data must translate into in order to be used by other programs.

本句中，in order to be used by other programs 作目的状语，修饰主句的谓语 involves。that is capable of reading the base or original language of the data, and identifying the language or languages that the data must translate into 是一个定语从句，修饰和限定 an executable program。在该从句中，and 连接了 is capable of 的两个宾语。that the data must translate into 是一个定语从句，修饰和限定 the language or languages。

Exercises

【Ex. 1】 根据课文内容回答问题。

1. What is data preprocessing?
2. What is data cleansing?
3. What does data cleansing involve at its most simple form?
4. What is data integration?
5. What are the two situations that call for data integration more than any other?
6. What is data transformation?
7. What is SQL?
8. What is the one difference between data transformation and data mediation?
9. What is data reduction?
10. When the data are already in digital form what do the "reduction'" of the data typically involve?

【Ex. 2】 把下列句子翻译为中文。

1. Effective solutions involve optimized techniques and technologies to extract, filter, and transform data.
2. The data on the replica may be inconsistent with the protected data so a consistency check is required.
3. This provides consistency and reduces the number of errors.
4. The class contains methods to insert, delete, and update a row or rowsfrom the database.
5. Examine the output to verify that all the commands were processedsuccessfully.
6. The backup device reported an unrecoverable hardware error.
7. Update statistics is run only on logged databases.
8. Reformat each record in the export file so that it can be used to modify each user account.
9. The report must include a dataset that specifies a connection to the package.

10. Setting up sharable, independent neutral information model is the precondition of information integration.

【Ex. 3】 短文翻译。

Data reduction is a term that applies to the business practice of accumulating, analyzing and ultimately transforming massive amounts of data into a series of summarized reports. The idea behind the data reduction process is to provide a complete though somewhat simplified format that can be utilized with relative ease in business settings. Several different approaches to the process may be used, with the selection of data reduction techniques and systems depending on the nature of the data and how those summary reports need to be structured in order to provide a full and comprehensive representation of that data.

One of the primary tasks in any type of data reduction effort is the organization of all data collected for the purpose. At times this portion of the process focuses on establishing some sort of order to the data that involves prioritizing in a consistent manner, using well-defined criteria to aid in the activity. Depending on the type of data involved, it is not unusual to include some rounding of certain figures in order to make the information easier to work with during the summarizing. Finally, the arrangement of the data into tables, columnar reports, or other types of labeling or formatting may be necessary in order to allow recipients of the reports to follow the logistics of the simplified information with relative ease.

【Ex. 4】 将下列词填入适当的位置（每词只用一次）。

sorting	reduction	images	processed	monitoring
nature	increase	diagnosis	transformation	encompasses

Data Reduction

Data reduction is the transformation of numerical or alphabetical digital information derived empirically or experimentally into a corrected, ordered, and simplified form. The basic concept is the __(1)__ of multitudinous amounts of data down to the meaningful parts.

When information is derived from instrument readings there may also be a __(2)__ from analog to digital form. When the data are already in digital form the "reduction" of the data typically involves some editing, scaling, coding, __(3)__, collating, and producing tabular summaries. When the observations are discrete but the underlying phenomenon is continuous then smoothing and interpolation are often needed. Often the data reduction is undertaken in the presence of reading or measurement errors. Some idea of the __(4)__ of these errors is needed before the most likely value may be determined.

An example in astronomy is the data reduction in the Kepler satellite. This satellite records 95-megapixel __(5)__ once every six seconds, generating tens of megabytes of data per second, which is orders of magnitudes more than the downlink bandwidth of 550 KBps. The on-board data reduction __(6)__ co-adding the raw frames for thirty minutes, reducing the bandwidth by a factor of 300. Furthermore, interesting targets are pre-selected and only the relevant pixels are processed, which is 6% of the total. This reduced data is then sent to Earth where it is __(7)__ further.

Research has also been carried out on the use of data reduction in wearable (wireless) devices for health __(8)__ and diagnosis applications. For example, in the context of epilepsy diagnosis, data reduction has been used to __(9)__ the battery lifetime of a wearable EEG device by selecting, and only transmitting, EEG data that is relevant for __(10)__ and discarding background activity.

Text B

Data Cleansing

Data cleansing or data cleaning is the process of detecting and correcting (or removing) corrupt or inaccurate records from a record set, table, or database. It refers to identifying incomplete, incorrect, inaccurate or irrelevant parts of the data and then replacing, modifying, or deleting the dirty or coarse data. Data cleansing may be performed interactively with data wrangling tools, or as batch processing through scripting.

After cleansing, a data set should be consistent with other similar data sets in the system. The inconsistencies detected or removed may have been originally caused by user entry errors, by corruption in transmission or storage, or by different data dictionary definitions of similar entities in different stores. Data cleansing differs from data validation in that validation almost invariably means data is rejected from the system at entry and is performed at the time of entry, rather than on batches of data.

The actual process of data cleansing may involve removing typographical errors or validating and correcting values against a known list of entities. The validation may be strict (such as rejecting any address that does not have a valid postal code) or fuzzy (such as correcting records that partially match existing, known records). Some data cleansing solutions will clean data by cross checking with a validated data set. A common data cleansing practice is data enhancement, where data is made more complete by adding related information. For example, appending addresses with any phone numbers related to that address. Data cleansing may also involve activities like harmonization of data and standardization of data. For example, harmonization of short codes (st, rd, etc.) to actual words (street, road, etcetera). Standardization of data is a means of changing a reference data set to a new standard, for example, use of standard codes.

1. Motivation

Administratively, incorrect or inconsistent data can lead to false conclusions and misdirected investments on both public and private scales. For instance, the government may want to analyze population census figures to decide which regions require further spending and investment on infrastructure and services. In this case, it will be important to have access to reliable data to avoid erroneous fiscal decisions. In the business world, incorrect data can be costly. Many companies use customer information databases that record data like contact information, addresses, and preferences. For instance, if the addresses are inconsistent, the company will suffer the cost of resending mail or even losing customers. The profession of forensic accounting and fraud investigating uses data cleansing in preparing its data and is typically done before data is sent to a data warehouse for further investigation. There are packages available so you can cleanse/wash address data while you enter it into your system. This is normally done via an application programming interface (API).

2. Data Quality

High-quality data needs to pass a set of quality criteria. Those include:

2.1 Validity

The degree to which the measures conform to defined business rules or constraints. When modern database technology is used to design data-capture systems, validity is fairly easy to ensure: invalid data arises mainly in legacy contexts (where constraints were not implemented in software) or where inappropriate data-capture technology was used (e.g., spreadsheets, where it is very hard to limit what a user chooses to enter into a cell, if cell validation is not used). Data constraints fall into the following categories:
- Data-type constraints—e.g., values in a particular column must be of a particular data type, e.g., Boolean, numeric (integer or real), date, etc.
- Range constraints: typically, numbers or dates should fall within a certain range. That is, they have minimum and/or maximum permissible values.
- Mandatory constraints: Certain columns cannot be empty.

- Unique Constraints: A field, or a combination of fields, must be unique across a dataset. For example, no two persons can have the same social security number.
- Set-membership constraints: The values for a column come from a set of discrete values or codes. For example, a person's gender may be Female, Male or Unknown (not recorded).
- Foreign-key constraints: This is the more general case of set membership. The set of values in a column is defined in a column of another table that contains unique values. For example, in a US taxpayer database, the "state" column is required to belong to one of the US's defined states or territories: the set of permissible states/territories is recorded in a separate States table. The term foreign key is borrowed from relational database terminology.
- Regular expression patterns: Occasionally, text fields will have to be validated this way. For example, phone numbers may be required to have the pattern (999) 999-9999.
- Cross-field validation: Certain conditions that utilize multiple fields must hold. For example, in laboratory medicine, the sum of the components of the differential white blood cell count must be equal to 100 (since they are all percentages). In a hospital database, a patient's date of discharge from hospital cannot be earlier than the date of admission.

2.2 Accuracy

The degree of conformity of a measure to a standard or a true value. Accuracy is very hard to achieve through data-cleansing in the general case, because it requires accessing an external source of data that contains the true value: such "gold standard" data is often unavailable. Accuracy has been achieved in some cleansing contexts, notably customer contact data, by using external databases that match up zip codes to geographical locations (city and state), and also help verify that street addresses within these zip codes actually exist.

2.3 Completeness

The degree to which all required measures are known. Incompleteness is almost impossible to fix with data cleansing methodology: one cannot infer facts that were not captured when the data in question was initially recorded. In some contexts, e.g., interview data, it may be possible to fix incompleteness by going back to the original source of data, i,e., re-interviewing the subject, but even this does not guarantee success because of problems of recall —e.g., in an interview to gather data on food consumption, no one is likely to remember exactly what one ate six months ago. In the case of systems that insist certain columns should not be empty, one may work around the problem by designating a value that indicates "unknown" or "missing", but supplying of default values does not imply that the data has been made complete.

2.4 Consistency

The degree to which a set of measures are equivalent in across systems. Inconsistency occurs when two data items in the data set contradict each other: e.g., a customer is recorded in two different systems with two different current addresses, and only one of them can be correct. Fixing inconsistency is not always possible: it requires a variety of strategies — e.g., deciding which data were recorded more recently, which data source is likely to be most reliable (the latter knowledge may be specific to a given organization), or simply trying to find the truth by testing both data items (e.g., calling up the customer).

2.5 Uniformity

The degree to which a set data measures are specified using the same units of measure in all systems. In datasets pooled from different locales, weight may be recorded either in pounds or kilos, and must be converted to a single measure using an arithmetic transformation.

The term integrity encompasses accuracy, consistency and some aspects of validation but is rarely used by itself in data-cleansing contexts because it is insufficiently specific. For example, "referential integrity" is a term used to refer to the enforcement of foreign-key constraints above.

3. Process

3.1 Data auditing

The data is audited using statistical and database methods to detect anomalies and contradictions: this eventually gives an indication of the characteristics of the anomalies and their locations. Several commercial software packages will let you specify constraints of various kinds (using a grammar that conforms to that of a standard programming language, e.g., JavaScript or Visual Basic) and then generate code that checks the data for violation of these constraints. For users who lack access to high-end cleansing software, Microcomputer database packages such as Microsoft Access or File Maker Pro will also let you perform such checks, on a constraint-by-constraint basis, interactively with little or no programming required in many cases.

3.2 Workflow specification

The detection and removal of anomalies is performed by a sequence of operations on the data known as the workflow. It is specified after the process of auditing the data and is crucial in achieving the end product of high-quality data. In order to achieve a proper workflow, the causes of the anomalies and errors in the data have to be closely considered.

3.3 Workflow execution

In this stage, the workflow is executed after its specification is complete and its correctness is verified. The implementation of the workflow should be efficient, even on large sets of data, which inevitably poses a trade-off because the execution of a data-cleansing operation can be computationally expensive.

3.4 Post-processing and controlling

After executing the cleansing workflow, the results are inspected to verify correctness. Data that could not be corrected during execution of the workflow is manually corrected, if possible. The result is a new cycle in the data-cleansing process where the data is audited again to allow the specification of an additional workflow to further cleanse the data by automatic processing.

Good quality source data has to do with "Data Quality Culture" and must be initiated at the top of the organization. It is not just a matter of implementing strong validation checks on input screens because almost no matter how strong these checks are, they can often still be circumvented by the users. There is a nine-step guide for organizations that wish to improve data quality:
- Declare a high level commitment to a data quality culture
- Drive process reengineering at the executive level
- Improve the data entry environment
- Improve application integration
- Change how processes work
- Promote end-to-end team awareness
- Promote interdepartmental cooperation
- Publicly celebrate data quality excellence
- Continuously measure and improve data quality

3.5 Parsing

A parser decides whether a string of data is acceptable within the allowed data specification. This is similar to the way a parser works with grammars and languages.

3.6 Data transformation

Data transformation allows the mapping of the data from its given format into the format expected by the appropriate application. This includes value conversions or translation functions, as well as normalizing numeric values to conform to minimum and maximum values.

3.7 Duplicate elimination

Duplicate detection requires an algorithm for determining whether data contains duplicate representations of the same entity. Usually, data is sorted by a key that would bring duplicate entries closer together for faster identification.

3.8 Statistical methods

By analyzing the data using the values of mean, standard deviation, range, or clustering algorithms, it is possible for an expert to find values that are unexpected and thus erroneous. Although the correction of such data is difficult since the true value is not known, it can be resolved by setting the values to an average or other statistical value. Statistical methods can also be used to handle missing values which can be replaced by one or more plausible values, which are usually obtained by extensive data augmentation algorithms.

4. System

The essential job of this system is to find a suitable balance between fixing dirty data and maintaining the data as close as possible to the original data from the source production system. This is a challenge for the extract, transform, load architect. The system should offer an architecture that can cleanse data, record quality events and measure/control quality of data in the data warehouse. A good start is to perform a thorough data profiling analysis that will help define the required complexity of the data cleansing system and also give an idea of the current data quality in the source systems.

5. Quality screens

Part of the data cleansing system is a set of diagnostic filters known as quality screens. Quality screens are divided into three categories:
- Column screens. Testing the individual column, e.g. for unexpected values like NULL values; non-numeric values that should be numeric; out of range values; etc.
- Structure screens. These are used to test for the integrity of different relationships between columns (typically foreign/primary keys) in the same or different tables.

They are also used for testing that a group of columns is valid according to some structural definition to which it should adhere.
- Business rule screens. The most complex of the three tests. They test to see if data, maybe across multiple tables, follow specific business rules. An example could be, that if a customer is marked as a certain type of customer, the business rules that define this kind of customer should be adhered to.

When a quality screen records an error, it can either stop the dataflow process, send the faulty data somewhere else than the target system or tag the data. The latter option is considered the best solution because the first option requires, that someone has to manually deal with the issue each time it occurs and the second implies that data are missing from the target system (integrity) and it is often unclear what should happen to these data.

6. Criticism of existing tools and processes

The main reasons cited are:
- Project costs: costs typically in the hundreds of thousands of dollars
- Time: lack of enough time to deal with large-scale data-cleansing software
- Security: concerns over sharing information, giving an application access across systems, and effects on legacy systems

7. Error event schema

The Error Event schema holds records of all error events thrown by the quality screens. It consists of an Error Event Fact table with foreign keys to three dimension tables that represent date (when), batch job (where) and screen (who produced error). It also holds information about exactly when the error occurred and the severity of the error. In addition there is an Error Event Detail Fact table with a foreign key to the main table that contains detailed information about in which table, record and field the error occurred and the error condition.

8. Challenges and problems

8.1 Error correction and loss of information

The most challenging problem within data cleansing remains the correction of values to remove duplicates and invalid entries. In many cases, the available information on such anomalies is limited and insufficient to determine the necessary transformations or corrections, leaving the deletion of such entries as a primary solution. The deletion of data, though, leads to loss of information; this loss can be particularly costly if there is a large amount of deleted data.

8.2 Maintenance of cleansed data

Data cleansing is an expensive and time-consuming process. So after having performed data cleansing and achieving a data collection free of errors, one would want to avoid the re-cleansing of data in its entirety after some values in data collection have changed. The process should only be repeated on values that have changed; this means that a cleansing lineage would need to be kept, which would require efficient data collection and management techniques.

8.3 Data cleansing in virtually integrated environments

In virtually integrated sources like IBM's DiscoveryLink, the cleansing of data has to be performed every time the data is accessed, which considerably increases the response time and lowers efficiency.

8.4 Data-cleansing framework

In many cases, it will not be possible to derive a complete data-cleansing graph to guide the process in advance. This makes data cleansing an iterative process involving significant exploration and interaction, which may require a framework in the form of a collection of methods for error detection and elimination in addition to data auditing. This can be integrated with other data-processing stages like integration and maintenance.

New Words

correct	[kə'rekt]	vt.改正，纠正
remove	[ri'mu:v]	vt.删除，移去，移动
corrupt	[kə'rʌpt]	adj.被破坏的
table	['teibl]	n.表，表格
incorrect	[ˌinkə'rekt]	adj.错误的，不正确的
irrelevant	[i'relivənt]	adj.不相关的
replace	[ri'pleis]	vt.取代，替换，代替
modify	['mɔdifai]	vt.更改，修改
coarse	[kɔ:s]	adj.粗糙的
inconsistency	[ˌinkən'sistənsi]	n.不一致，矛盾
invariably	[in'veəriəbli]	adv.不变地，总是

harmonization	[,hɑ:mənai'zeiʃən]	n.一致，融洽
standardization	[,stændədai'zeiʃən]	n.标准化
infrastructure	['infrəstrʌktʃə]	n.基础设施
erroneous	[i'rəuniəs]	adj.错误的，不正确的
investigation	[in,vesti'geiʃən]	n.调查，研究
inappropriate	[,inə'prəupriət]	adj.不适当的，不相称的
cell	[sel]	n.单元
real	['ri:əl]	adj.实际的
minimum	['miniməm]	adj.最小的，最低的 n.最小值，最小化
maximum	['mæksiməm]	n.最大量，最大限度 adj.最大极限的
mandatory	['mændətəri]	adj.命令的，强制的
taxpayer	['tæks,peiə]	n.纳税人
borrowed	['bɔrəud]	adj.借来的
conformity	[kən'fɔ:miti]	n.一致，符合
unavailable	[,ʌnə'veiləbl]	adj.难以获得的
contradict	[kɔntrə'dikt]	vt.同……矛盾，同……抵触
reliable	[ri'laiəbl]	adj.可靠的，可信赖的
uniformity	[,ju:ni'fɔ:miti]	n.均匀性
insufficiently	[,insə'fiʃəntli]	adv.不够地，不能胜任地
contradiction	[,kɔntrə'dikʃən]	n.矛盾，反驳
indication	[,indi'keiʃən]	n.指出，指示，迹象，暗示
violation	[,vaiə'leiʃən]	n.违反，违背，妨碍
microcomputer	['maikrəukəmpju:tə]	n.微型电子计算机
interactively	[,intər'æktivli]	adv.交互式地
sequence	['si:kwəns]	n.次序，顺序，序列
correctness	[kə'rektnəs]	n.正确性
inevitably	[in'evitəbli]	adv.不可避免地
pose	[pəuz]	v.形成，引起，造成
trade-off	[treid-ɔ:f]	n.交换，平衡
inspect	[in'spekt]	v.检查
automatic	[,ɔ:tə'mætik]	adj.自动的 n.自动机械
interdepartmental	[,intədi,pɑ:t'mentl]	adj.各部门间的
parse	[pɑ:z]	vt.解析，分解
parser	['pɑ:sə]	n.解析器，解释器

normalize	[ˈnɔːməlaiz]	v.规格化
duplicate	[ˈdjuːplikit]	adj.重复的
elimination	[iˌlimiˈneiʃən]	n.排除，除去，消除，消灭
unexpected	[ˌʌniksˈpektid]	adj.想不到的，意外的，未预料到的
plausible	[ˈplɔːzəbl]	adj.似是而非的
augmentation	[ˌɔːgmenˈteiʃən]	n.增加，增强
cleanse	[klenz]	v.净化，提纯
screen	[skriːn]	vt.筛选
diagnostic	[ˌdaiəgˈnɔstik]	adj.诊断的
follow	[ˈfɔləu]	vt.遵循
severity	[siˈveriti]	n.严格，严重
maintenance	[ˈmeintinəns]	n.维护，保持
iterative	[ˈitərətiv]	adj.重复的，反复的；迭代的 n.反复体，循环体
exploration	[ˌeksplɔːˈreiʃən]	n.探测，探查

📖 Phrases

data cleansing	数据清理
record set	记录集
coarse data	粗糙数据
data wrangling	数据整理
data dictionary	数据字典
typographical error	排字错误，误排
population census	人口普查
forensic accounting	法律财会专业；法务会计学
fraud investigating	欺诈调查，舞弊调查
data type	数据类型
maximum permissible value	最大容许值
social security number	社会保险号码
discrete value	离散值，不连续值
match up	使调协，使配合
referential Integrity	参照完整性
data auditing	数据审核，数据审计
conform to	符合，遵照
in order to …	为了……
no matter how	不管如何

missing value	遗漏值，漏测值
extensive data augmentation algorithm	扩展数据增强算法
primary key	主键
foreign key	外键
error correction	纠错，数据纠正
free of	无……的，摆脱……的
management techniques	管理技术，管理方法
response time	响应时间

Exercises

【Ex. 5】根据课文内容回答问题。

1. What is data cleansing ?
2. Why does data cleansing differ from data validation?
3. What can lead to false conclusions and misdirected investments on both public and private scales?
4. What does high-quality data need to pass?
5. Why is accuracy is very hard to achieve through data-cleansing in the general case?
6. Why is good quality source data not just a matter of implementing strong validation checks on input screens?
7. What is the essential job of this system?
8. How many categories are quality screens divided into? What are they?
9. What are the main reasons cited for criticism of existing tools and processes?
10. What is the most challenging problem within data cleansing?

参考译文

数据预处理

　　数据预处理是一种数据挖掘技术，它将原始数据转换为可理解的格式。真实世界的数据通常不完整、不一致和/或缺乏某些行为或趋势，并且可能包含许多错误。数据预处理是解决此类问题的经过检验的方法。数据预处理准备原始数据以供进一步处理。数据在预处理过程中会经过一系列的步骤。

1. 数据清理

数据清理（也称为数据清洗）是确保一组数据正确及准确的过程。在此过程中，检查记录的准确性和一致性，并根据需要进行更正或删除。这可能发生在一组记录中，也可能发生在需要合并或将协同工作的多组数据之间。

最简单的数据清理形式是一个人或一些人阅读一组记录并校验其准确性。纠正错字和拼写错误、修正标记错了的数据以及归档并完善不完整或缺失的条目。这些操作通常会清除过期或不可恢复的记录，以免它们占用空间并降低操作效率。

更复杂的数据清理操作可由计算机程序执行。这些程序可以根据用户确定的各种规则和程序来检查数据。可以设置一个程序来删除在过去五年内未更新的所有记录，更正任何有拼写错误的单词，并删除全部重复的副本。更复杂的程序可能会根据正确的邮政编码填写一个缺失的城市，或者将数据库中所有项目的价格更改为以其他类型货币的计价。

2. 数据整合

数据整合是将多个数据源合并成单个数据源。这种操作往往非常耗费时间，因为不同的数据源可能彼此不兼容。例如像电子表格中的不同列名称这样简单的事情就足以要求重新格式化日期。在两个组织刚开始没有联网、已经独立工作后才联网的情况下，这个过程很常见。由于免费数据源和在线数据库的普及，数据整合已经成为一个更重要的课题。

只要存储在计算机系统中的都可以是数据整合的数据部分。数据的实际内容通常没有存储数据的方式那样重要。大多数情况下，数据保存在有组织的信息系统的数据库中。这些系统包含唯一的条目和字段，允许用户快速查找信息。

任何数据整合过程的最大障碍就是数据本身。在许多情况下，当数据首次建立时，并未考虑将数据集与另一个数据集合并。这意味着即使两个数据集可能是相同的东西，它们也完全不兼容。

几乎任何东西都会使数据库不兼容。简单的如不同演示文稿，其字段顺序或列宽就足以阻止二者轻松的合并。当数据差异巨大时（例如包含更多或更少信息的一个数据库）合并要困难得多。

在商业和研究这两个领域要求数据整合的呼声比任何其他领域都强烈。在商业界，合并部门或公司数据需要将以前单独的信息组合成单一的结构。这种整合通常非常困难，除非原始组织使用了类似的软件并具有类似的信息目标。

当为研究目的进行数据整合时，通常会更加顺畅。当一位研究人员向他人提供信息时，双方通常会研究相同的过程。这意味着他们将使用类似的方法对其数据进行分类和存储。

过去数据整合是数据研究的一个相对较小的领域，自 21 世纪初以来，情况发生了变化。随着免费在线数据库变得越来越流行和准确，公司正在争取以可共享的格式获取他们的信息。这允许他们以公共形式发布其信息，并将一些著名的公共接口的私有版本集成到自己的系统中。

3. 数据转换

数据转换是将信息或数据从一种格式转换成另一格式的过程。虽然通常的策略都是将文档从一种格式转换为另一种格式，但数据转换也可把一种计算机语言编写的程序转换为另一种语言编写的程序，这样便于程序在特定平台上运行。实际的转换可能将多个数据流转换成通用格式，或者将单个格式转换成多个不同的形式，以便在广泛的平台上使用。

数据转换的过程涉及使用所谓的 SQL 或结构化查询语言。SQL 是一种计算机语言，它负责管理存储在某种类型的数据管理系统中的信息。

在实际使用中，数据转换使用可执行程序，该程序能够读取基础数据或原始语言，并且识别该语言，并将其转换为其他程序可以使用的数据。一旦完成了转换的映射，程序就将数据转换为所需的单个或多个格式，并相应地发布转换的数据。在许多应用中，这在几秒钟内就可完成。

一个类似的过程被称为数据调节。像数据转换一样，该想法是使一种格式的数据能够以另一种格式使用。与数据调节不同，数据映射过程涉及创建所谓的数据模型，作为所涉及的两种格式之间的中介，而不是直接转换信息。

与许多类型的计算机技术一样，数据转换也在不断发展，因为新程序有助于提高信息转换效率和扩大转换范围。随着这个过程中包含的程序和格式越来越多，在许多不同平台上完全不兼容的数据得以共享。在全球设置中，协作者并不总是使用相同程序或语言作为数据系统的基础，这些持续改进意味着在系统之间手工转换和输入数据的时间更少了。

4. 数据缩减

数据缩减是将从经验或实验中得出的数字或字母数字信息转换为正确、有序和简化的形式。基本概念是将大量数据减少到有意义的部分。

当信息来源于仪器读数时，也可能会出现从模拟形式到数字形式的变化。当数据已经是数字形式时，数据的"减少"通常涉及一些编辑、缩放、编码、排序、整理和生成表格摘要。如果观察结果是离散的，但是潜在的现象是连续的，则通常需要进行平滑和插值处理。通常要在读取或测量错误的情况下进行数据缩减。在确定最有可能的价值之前，需要考虑这些错误的性质。

  一个天文学的例子是开普勒卫星的数据缩减。该卫星每六秒记录一次 95 万像素的图像，每秒生成数十兆字节的数据，这大于 550KBps 的下行链路带宽的数量级。在轨数据减少包括合并了三十分钟的原始数据，带宽减少到原来的 1/300。此外，预先选择感兴趣的目标，并只处理相关像素，这只占总数的 6%。然后将此减少了的数据发送到地球，进一步处理。

  在可穿戴（无线）装置中，健康监测和诊断应用也使用数据缩减。例如，在癫痫诊断时，通过选择并且仅发送与诊断相关的 EEG 数据和丢弃背景活动数据，使用数据缩减来增加可戴式 EEG 设备的电池寿命。

Unit 9

Text A

Data Mining

Data mining is a powerful new technology with great potential to help companies focus on the most important information in the data they have collected about the behavior of their customers and potential customers. It discovers information within the data that queries and reports can't effectively reveal.

1. What is Data Mining?

Data mining, or knowledge discovery, is the computer-assisted process of digging through and analyzing enormous sets of data and then extracting the meaning of the data. Data mining tools predict behaviors and future trends, allowing businesses to make proactive, knowledge-driven decisions. Data mining tools can answer business questions that traditionally were too time-consuming to resolve. They scour databases for hidden patterns, finding predictive information that experts may miss because it lies outside their expectations.

Data mining derives its name from the similarities between searching for valuable information in a large database and mining a mountain for a vein of valuable ore. Both processes require either sifting through an immense amount of material, or intelligently probing it to find where the value resides.

2. What Can Data Mining Do?

Although data mining is still in its infancy, companies in a wide range of industries — including retail, finance, heath care, manufacturing transportation, and aerospace — are already using data mining tools and techniques to take advantage of historical data. By using pattern recognition technologies and statistical and mathematical techniques to sift through warehoused information, data mining helps analysts recognize significant facts, relationships, trends, patterns, exceptions and anomalies that might otherwise go unnoticed.

For businesses, data mining is used to discover patterns and relationships in the data in order to help make better business decisions. Data mining can help spot sales trends, develop smarter marketing campaigns, and accurately predict customer loyalty. Specific uses of data mining include:

- Market segmentation—Identify the common characteristics of customers who buy the same products from your company.
- Customer churn—Predict which customers are likely to leave your company and go to a competitor.
- Fraud detection—Identify which transactions are most likely to be fraudulent.
- Direct marketing—Identify which prospects should be included in a mailing list to obtain the highest response rate.
- Interactive marketing—Predict what each individual accessing a Web site is most likely interested in seeing.
- Market basket analysis—Understand what products or services are commonly purchased together; e.g., beer and diapers.
- Trend analysis—Reveal the difference between a typical customer this month and last.

Data mining technology can generate new business opportunities by:

Automated prediction of trends and behaviors: Data mining automates the process of finding predictive information in a large database. Questions that traditionally required extensive hands-on analysis can now be directly answered from the data. A typical example of a predictive problem is targeted marketing. Data mining uses data on past promotional mailings to identify the targets most likely to maximize return on investment in future mailings. Other predictive problems include forecasting bankruptcy and other forms of default, and identifying segments of a population likely to respond similarly to given events.

Automated discovery of previously unknown patterns: Data mining tools sweep through databases and identify previously hidden patterns. An example of pattern discovery is the analysis of retail sales data to identify seemingly unrelated products that are often purchased together. Other pattern discovery problems include detecting fraudulent credit card transactions and identifying anomalous data that could represent data entry keying errors.

Using massively parallel computers, companies dig through volumes of data to discover patterns about their customers and products. For example, grocery chains have found that when men go to a supermarket to buy diapers, they sometimes walk out with a six-pack of beer as well. Using that information, it's possible to lay out a store so that these items are closer.

AT&T, A.C. Nielson, and American Express are among the growing ranks of companies implementing data mining techniques for sales and marketing. These systems are crunching through terabytes of point-of-sale data to aid analysts in understanding consumer behavior and promotional strategies. Why? To gain a competitive advantage and increase profitability!

Similarly, financial analysts are plowing through vast sets of financial records, data feeds, and other information sources in order to make investment decisions. Health-care organizations are examining medical records to understand trends of the past so they can reduce costs in the future.

3. How Data Mining Works?

How is data mining able to tell you important things that you didn't know or what is going to happen next? The technique that is used to perform these feats is called modeling. Modeling is simply the act of building a model (a set of examples or a mathematical relationship) based on data from situations where the answer is known and then applying the model to other situations where the answers aren't known. Modeling techniques have been around for centuries, of course, but it is only recently that data storage and communication capabilities required to collect and store huge amounts of data, and the computational power to automate modeling techniques to work directly on the data have been available.

As a simple example of building a model, consider the director of marketing for a telecommunications company. He would like to focus his marketing and sales efforts on segments of the population most likely to become big users of long distance services. He knows a lot about his customers, but it is impossible to discern the common characteristics of his best customers because there are so many variables. From his existing database of customers, which contains information such as age, sex, credit history, income, zip code, occupation, etc., he can use data mining tools, such as neural networks, to identify the characteristics of those customers who make lots of long distance calls. For instance, he might learn that his best customers are unmarried females between the age of 34 and 42 who make in excess of $60,000 per year. This, then, is his model for high value customers, and he would budget his marketing efforts accordingly.

4. Data Mining Technologies

The analytical techniques used in data mining are often well-known mathematical algorithms and techniques. What is new is the application of those techniques to general business problems made possible by the increased availability of data and inexpensive storage and processing power. Also, the use of graphical interfaces has led to tools becoming available that business experts can easily use.

Some of the tools used for data mining are:

Artificial neural networks—Non-linear predictive models that learn through training and resemble biological neural networks in structure.

Decision trees—Tree-shaped structures that represent sets of decisions. These decisions generate rules for the classification of a dataset.

Rule induction—The extraction of useful if-then rules from data based on statistical significance.

Genetic algorithms — Optimization techniques based on the concepts of genetic combination, mutation, and natural selection.

Nearest neighbor—A classification technique that classifies each record based on the records most similar to it in an historical database.

5. Real-World Examples

Details about who calls whom, how long they are on the phone, and whether a line is used for fax as well as voice can be invaluable in targeting sales of services and equipment to specific customers. But these tidbits are buried in masses of numbers in the database. By delving into its extensive customer-call database to manage its communications network, a regional telephone company identifies new types of unmet customer needs. Using its data mining system, it discovers how to pinpoint prospects for additional services by measuring daily household usage for selected periods. For example, households that make many lengthy calls between 3 p.m. and 6 p.m. are likely to include teenagers who are prime candidates for their own phones and lines. When the company uses target marketing that emphasizes convenience and value for adults—"Is the phone always tied up?"—hidden demand surfaces. Extensive telephone use between 9 a.m. and 5 p.m. characterized by patterns related to voice, fax, and modem usage suggests a customer has business activity. Target marketing offering those customers "business communications capabilities for small budgets" results in sales of additional lines, functions, and equipment.

The ability to accurately gauge customer response to changes in business rules is a powerful competitive advantage. A bank searching for new ways to increase revenues from its credit card operations tested a nonintuitive possibility: Would credit card usage and interest earned increase significantly if the bank halved its minimum required payment? With hundreds of gigabytes of data representing two years of average credit card balances, payment amounts, payment timeliness, credit limit usage, and other key parameters, the bank used a powerful data mining system to model the impact of the proposed policy change on specific customer categories. The bank discovered that cutting minimum payment requirements for small, targeted customer categories could increase average balances and extend indebtedness periods, generating more than $25 million in additional interest earned. Merck-Medco Managed Care is a mail-order business which sells drugs to the country's largest health care providers. Merck-Medco is mining its one terabyte data warehouse to uncover hidden links between illnesses and known drug treatments, and spot trends that help pinpoint which drugs are the most effective for what types of patients. The results are more effective treatments that are also less costly. Merck-Medco's data mining project has helped customers save an average of 10%-15% on prescription costs.

6. The Future of Data Mining

In the short-term, the results of data mining will be in profitable business related areas. Micro-marketing campaigns will explore new niches. Advertising will target potential customers with new precision.

In the medium term, data mining may be as common and easy to use as e-mail. We may use these tools to find the best airfare to New York, root out a phone number of a long-lost classmate, or find the best prices on lawn mowers.

The long-term prospects are truly exciting. Imagine intelligent agents turning loose on medical research data or on sub-atomic particle data. Computers may reveal new treatments for diseases.

New Words

behavior	[bi'heiviə]	n.举止，行为
discover	[dis'kʌvə]	vt.发现，发觉
dig	['dig]	v.掘，挖，搜集
proactive	[ˌprəu'æktiv]	adj.积极的，主动地
time-consuming	['taimkənˌsju:miŋ]	adj.耗费时间的，旷日持久的
scour	['skauə]	v.四处搜集，冲洗，擦亮
expectation	[ˌekspek'teiʃən]	n.期待，预料，指望，展望
similarity	[ˌsimi'læriti]	n.类似，类似处

vein	[vein]	n.矿脉，纹理
probe	['prəub]	v.探查，探测
transportation	[,trænspɔː'teiʃən]	n.运输，运送
aerospace	['ɛərəuspeis]	n.航空航天
sift	[sift]	vt.筛分，精选；审查
		vi.筛；细查
relationship	[ri'leiʃənʃip]	n.关系，关联
anomaly	[ə'nɔməli]	n.不规则，异常的人或物
unnoticed	[ʌn'nəutist]	adj.不引人注意的，被忽视的
spot	[spɔt]	vt.认出，发现
segmentation	[,segmən'teiʃən]	n.分割
churn	[tʃəːn]	v.流失
fraudulent	['frɔːdjulənt]	adj.欺诈的，欺骗性的
bankruptcy	['bæŋkrəpsi] ['bæŋkrʌptsi]	n.破产
sweep	[swiːp]	v.扫过，掠过
seemingly	['siːmiŋli]	adv.表面上地
anomalous	[ə'nɔmələs]	adj.不规则的，反常的
grocery	['grəusəri]	n.食品杂货店，食品店，杂货铺
crunch	[krʌntʃ]	v.嘎吱嘎吱地咬嚼，压碎，嘎吱嘎吱地踏过
feat	[fiːt]	n.技艺；功绩，壮举
discern	[di'səːn]	v.认识，洞悉，辨别，看清楚
occupation	[,ɔkju'peiʃən]	n.职业
budget	['bʌdʒit]	n.预算
		vi.做预算，编入预算
inexpensive	[,iniks'pensiv]	adj.便宜的，不贵重的
artificial	[,ɑːti'fiʃəl]	adj.人造的
non-linear	[nɔn-'liniə]	adj.非线性的
induction	[in'dʌkʃən]	n.归纳
optimization	[,ɔptimai'zeiʃən]	n.最佳化，最优化
mutation	[mjuː'teiʃən]	n.变化，转变；（生物物种的）突变
invaluable	[in'væljuəbl]	adj.无价的，价值无法衡量的
tidbit	['tidbit]	n.一小口（食物），花絮
bury	['beri]	vt.掩埋，隐藏
unmet	[ʌn'met]	adj.未满足的，未相遇的，未应付的
pinpoint	['pinpɔint]	n.精确

		*adj.*极微小的
		*v.*查明
nonintuitive	[nɔnin'tjuːitiv]	*adj.*非直觉的
possibility	[ˌpɔsi'biliti]	*n.*可能性
earn	[əːn]	*v.*赚得，获得
halve	[hɑːv]	*vt.*二等分，平分，分享，减半
indebtedness	[in'detidnis]	*n.*亏欠，债务
mail-order	[meil-'ɔːdə]	*adj.*邮购的
uncover	[ʌn'kʌvə]	*v.*揭示
drug	[drʌg]	*n.*药，麻药
		*vi.*吸毒
		*vt.*使服毒品，毒化
treatment	['triːtmənt]	*n.*处理，治疗
prescription	[pri'skripʃən]	*n.*处方，药方
profitable	['prɔfitəbl]	*adj.*有利可图的
niche	[nitʃ]	*n.*小生态环境，商机

🖎 Phrases

knowledge discovery	知识发现
computer-assisted process	计算机辅助过程
knowledge-driven decision	知识驱动决策
sift through	筛选
heath care	卫生保健
pattern recognition	模式识别
business decision	业务决策，商务决定
customer churn	客户流失
response rate	响应率
market basket analysis	购物篮分析
business opportunity	业务机会，商业机会
targeted marketing	目标市场
seemingly unrelated product	看上去无关的产品
point-of-sale data	销售终端数据
investment decision	投资决策
neural network	神经网络
graphical interface	图形界面，图形接口
predictive model	预测模型

decision tree	决策树
rule induction	规则归纳
genetic algorithm	遗传算法
nearest neighbor	最邻近算法
delve into	钻研，深入研究
root out	搜寻
lawn mower	割草机，剪草机

Notes

[1] Data mining is a powerful new technology with great potential to help companies focus on the most important information in the data they have collected about the behavior of their customers and potential customers.

本句中，to help companies focus on the most important information in the data they have collected about the behavior of their customers and potential customers 是一个动词不定式短语，做定语，修饰和限定 potential。在该不定式短语中，they have collected about the behavior of their customers and potential customers 是一个定语从句，修饰和限定 data。focus on 的意思是"注重，关注"。

[2] An example of pattern discovery is the analysis of retail sales data to identify seemingly unrelated products that are often purchased together.

本句中，the analysis of retail sales data to identify seemingly unrelated products that are often purchased together 是一个名词性短语，作表语。在该名词短语中，to identify seemingly unrelated products that are often purchased together 是一个动词不定式短语，作定语，修饰和限定 retail sales data。在该不定式短语中，that are often purchased together 是一个定语从句，修饰和限定 products。

[3] Modeling is simply the act of building a model (a set of examples or a mathematical relationship) based on data from situations where the answer is known and then applying the model to other situations where the answers aren't known.

本句中，where the answer is known 是一个定语从句，修饰和限定 situations。where the answers aren't known 也是一个定语从句，修饰和限定 other situations。based on 的意思是"基于，根据"；apply... to 的意思是"将……应用于"。

[4] From his existing database of customers, which contains information such as age, sex, credit history, income, zip code, occupation, etc., he can use data mining tools, such as neural networks, to identify the characteristics of those customers who make lots of long distance calls.

本句中，which contains information such as age, sex, credit history, income, zip code, occupation, etc.是一个非限定性定语从句，对 his existing database of customers 进行补

充说明。to identify the characteristics of those customers who make lots of long distance calls 是一个动词不定式短语，作目的状语，修饰主句的谓语 use。在该不定式短语中，who make lots of long distance calls 是一个定语从句，修饰和限定 those customers。

Exercises

【Ex.1】 根据课文内容回答问题。

1. What is data mining?
2. Where does data mining derive its name from?
3. What does data mining help analysts do？And how?
4. What are the specific uses of data mining mentioned in the passage?
5. What is a typical example of a predictive problem? What is an example of pattern discovery?
6. Why are financial analysts are plowing through vast sets of financial records, data feeds, and other information sources? Why are health-care organizations examining medical records?
7. What is modeling?
8. What are some of the tools used for data mining?
9. What is the result of Merck-Medco's data mining project?
10. What is the future of data mining?

【Ex.2】根据给出的汉语词义和规定的词类写出相应的英语单词。每词的首字母已给出。

v.掘，挖，搜集	d
n.期待，预料，指望，展望	e
n.运输，运送	t
n.建模	m
n.归纳	i
n.算法	a
n.小生态环境，商机	n
n.精确	p
adj.非线性的	n
n.最佳化，最优化	o
adj.计算的	c
v.流失	c
vt.筛分，精选；审查	s

n.关系，关联	r _____
v.探查，探测	p _____
vt.发现，发觉	d _____
vt.认出，发现	s _____
n.处理，治疗	t _____
n.可能性	p _____
n.抽出，取出	e _____

【Ex.3】把下列句子翻译为中文。

1. They discovered how to form the image in a thin layer on the surface.
2. They will probe deeply into the matter.
3. Campuses are usually accessible by public transportation.
4. Segmentation of the market allows the bank to tailor its approach to the customers' requirement.
5. What modeling program are you using (include version number)?
6. Students may pursue research in any aspect of computational linguistics.
7. He wanted to look for occupation suited to his abilities.
8. There has been an underspend in the department's budget.
9. As to sequential pattern mining, mining algorithm is very important.
10. The dataset must have a table before a relationship can be added.

【Ex.4】将下列词填入适当的位置（每词只用一次）。

match	employees	campaigns	build	relationships
improve	sufficient	requirements	marketing	maximizing

CRM (customer relationship management) is an information industry term for methodologies, software, and usually Internet capabilities that help an enterprise manage customer relationships in an organized way. For example, an enterprise might __(1)__ a database about its customers that described relationships in __(2)__ detail so that management, salespeople, people providing service, and perhaps the customer could directly access information, __(3)__ customer needs with product plans and offerings, remind customers of service __(4)__, know what other products a customer had purchased, and so forth.

According to one industry view, CRM consists of:
- Helping an enterprise to enable its __(5)__ departments to identify and target their best customers, manage marketing __(6)__ and generate quality leads for the sales

team.
- Assisting the organization to __(7)__ telesales, account, and sales management by optimizing information shared by multiple employees, and streamlining existing processes (for example, taking orders using mobile devices).
- Allowing the formation of individualized relationships with customers, with the aim of improving customer satisfaction and __(8)__ profits; identifying the most profitable customers and providing them the highest level of service.
- Providing __(9)__ with the information and processes necessary to know their customers, understand and identify customer needs and effectively build __(10)__ between the company, its customer base, and distribution partners.

Text B

Top 6 Data Mining Algorithms

1. C4.5

What does it do? C4.5 constructs a classifier in the form of a decision tree. In order to do this, C4.5 is given a set of data representing things that are already classified.

Wait, what's a classifier? A classifier is a tool in data mining that takes a bunch of data representing things we want to classify and attempts to predict which class the new data belongs to.

What's an example of this? Sure, suppose a dataset contains a bunch of patients. We know various things about each patient like age, pulse, blood pressure, family history, etc. These are called attributes.

Now:

Given these attributes, we want to predict whether the patient will get cancer. The patient can fall into 1 of 2 classes: will get cancer or won't get cancer. C4.5 is told the class for each patient.

And here's the deal:

Using a set of patient attributes and the patient's corresponding class, C4.5 constructs a decision tree that can predict the class for new patients based on their attributes.

Cool, so what's a decision tree? Decision tree learning creates something similar to a flowchart to classify new data. Using the same patient example, one particular path in the flowchart could be:

(1) Patient has a history of cancer

(2) Patient is expressing a gene highly correlated with cancer patients

(3) Patient has tumors

(4) Patient's tumor size is greater than 5cm

The bottom line is:

At each point in the flowchart is a question about the value of some attribute, and depending on those values, he or she gets classified.

Is this supervised or unsupervised? This is supervised learning, since the training dataset is labeled with classes. Using the patient example, C4.5 doesn't learn on its own that a patient will get cancer or won't get cancer. We told it first, it generated a decision tree, and now it uses the decision tree to classify.

You might be wondering how C4.5 is different from other decision tree systems?

(1) First, C4.5 uses information gain when generating the decision tree.

(2) Second, although other systems also incorporate pruning, C4.5 uses a single-pass pruning process to mitigate over-fitting. Pruning results in many improvements.

(3) Third, C4.5 can work with both continuous and discrete data. My understanding is it does this by specifying ranges or thresholds for continuous data thus turning continuous data into discrete data.

(4) Finally, incomplete data is dealt with in its own ways.

Why use C4.5? Arguably, the best selling point of decision trees is their ease of interpretation and explanation. They are also quite fast, quite popular and the output is human readable.

Where is it used? A popular open-source Java implementation can be found over at OpenTox.Orange, an open-source data visualization and analysis tool for data mining, which implements C4.5 in their decision tree classifier.

Classifiers are great, but make sure to checkout the next algorithm about clustering…

2. k-means

What does it do? k-means creates k groups from a set of objects so that the members of a group are more similar. It's a popular cluster analysis technique for exploring a dataset.

Hang on, what's cluster analysis? Cluster analysis is a family of algorithms designed to form groups such that the group members are more similar versus non-group members. Clusters and groups are synonymous in the world of cluster analysis.

Is there an example of this? Definitely, suppose we have a dataset of patients. In cluster analysis, these would be called observations. We know various things about each patient like age, pulse, blood pressure, etc. This is a vector representing the patient.

Look:

You can basically think of a vector as a list of numbers we know about the patient. This list can also be interpreted as coordinates in multi-dimensional space. Pulse can be one dimension, blood pressure another dimension and so forth.

You might be wondering:

Given this set of vectors, how do we cluster together patients that have similar age, pulse, blood pressure, etc?

Want to know the best part?

You tell k-means how many clusters you want. k-means takes care of the rest.

How does k-means take care of the rest? k-means has lots of variations to optimize for certain types of data.

At a high level, they all do something like this:

(1) k-means picks points in multi-dimensional space to represent each of the k clusters. These are called centroids.

(2) Every patient will be closest to 1 of these k centroids. They hopefully won't all be closest to the same one, so they'll form a cluster around their nearest centroid.

(3) What we have are k clusters, and each patient is now a member of a cluster.

(4) k-means then finds the center for each of the k clusters based on its cluster members (yep, using the patient vectors!).

(5) This center becomes the new centroid for the cluster.

(6) Since the centroid is in a different place now, patients might now be closer to other centroids. In other words, they may change cluster membership.

(7) Steps(2)-(6) are repeated until the centroids no longer change, and the cluster memberships stabilize. This is called convergence.

Is this supervised or unsupervised? It depends, but most would classify k-means as unsupervised. Other than specifying the number of clusters, k-means "learns" the clusters on its own without any information about which cluster an observation belongs to. k-means can be semi-supervised.

Why use k-means? I don't think many will have an issue with this:

The key selling point of k-means is its simplicity. Its simplicity means it's generally faster and more efficient than other algorithms, especially over large datasets.

It gets better:

k-means can be used to pre-cluster a massive dataset followed by a more expensive cluster analysis on the sub-clusters. k-means can also be used to rapidly "play" with k and explore whether there are overlooked patterns or relationships in the dataset.

It's not all smooth sailing:

Two key weaknesses of k-means are its sensitivity to outliers, and its sensitivity to the initial choice of centroids. One final thing to keep in mind is k-means is designed to operate on continuous data—you'll need to do some tricks to get it to work on discrete data.

Where is it used? A ton of implementations for k-means clustering are available online: Apache Mahout, Julia, R, SciPy, Weka, MATLAB, SAS.

If decision trees and clustering didn't impress you, you're going to love the next algorithm.

3. Support vector machines

What does it do? Support vector machine (SVM) learns a hyperplane to classify data into 2 classes. At a high-level, SVM performs a similar task like C4.5 except SVM doesn't use decision trees at all.

As it turns out…

SVM can perform a trick to project your data into higher dimensions. Once projected into higher dimensions…

…SVM figures out the best hyperplane which separates your data into the 2 classes.

Do you have an example? Absolutely, the simplest example I found starts with a bunch of red and blue balls on a table. If the balls aren't too mixed together, you could take a stick and without moving the balls, separate them with the stick.

You see:

When a new ball is added on the table, by knowing which side of the stick the ball is on, you can predict its color.

What do the balls, table and stick represent? The balls represent data points, and the red and blue color represent 2 classes. The stick represents the simplest hyperplane which is a line.

And the coolest part?

SVM figures out the function for the hyperplane.

What if things get more complicated? Right, they frequently do. If the balls are mixed together, a straight stick won't work.

Here's the work-around:

Quickly lift up the table throwing the balls in the air. While the balls are in the air and thrown up in just the right way, you use a large sheet of paper to divide the balls in the air.

You might be wondering if this is cheating:

Nope, lifting up the table is the equivalent of mapping your data into higher dimensions. In this case, we go from the 2 dimensional table surface to the 3 dimensional balls in the air.

How does SVM do this? By using a kernel we have a nice way to operate in higher dimensions. The large sheet of paper is still called a hyperplane, but it is now a function for a plane rather than a line.

How do balls on a table or in the air map to real-life data? A ball on a table has a location that we can specify using coordinates. For example, a ball could be 20cm from the left edge and 50cm from the bottom edge. Another way to describe the ball is as (x, y) coordinates or (20, 50). x and y are 2 dimensions of the ball.

Here's the deal:

If we had a patient dataset, each patient could be described by various measurements like pulse, blood pressure, etc. Each of these measurements is a dimension.

The bottom line is:

SVM does its thing, maps them into a higher dimension and then finds the hyperplane to separate the classes.

Margins are often associated with SVM? What are they? The margin is the distance between the hyperplane and the 2 closest data points from each respective class. In the ball and table example, the distance between the stick and the closest red and blue ball is the margin.

The key is:

SVM attempts to maximize the margin, so that the hyperplane is just as far away from red ball as the blue ball. In this way, it decreases the chance of misclassification.

Where does SVM get its name from? Using the ball and table example, the hyperplane is equidistant from a red ball and a blue ball. These balls or data points are called support vectors, because they support the hyperplane.

Is this supervised or unsupervised? This is a supervised learning, since a dataset is used to first teach the SVM about the classes. Only then is the SVM capable of classifying new data.

Why use SVM? SVM along with C4.5 are generally the 2 classifiers to try first. No classifier will be the best in all cases due to the No Free Lunch Theorem. In addition, kernel selection and interpretability are some weaknesses.

Where is it used? There are many implementations of SVM. A few of the popular ones are scikit-learn, MATLAB and of course libsvm.

4. Apriori

The Apriori algorithm learns association rules and is applied to a database containing a large number of transactions.

What are association rules? Association rule learning is a data mining technique for learning correlations and relations among variables in a database.

What's an example of Apriori? Let's say we have a database full of supermarket transactions. You can think of a database as a giant spreadsheet where each row is a customer transaction and every column represents a different grocery item.

Here's the best part:

By applying the Apriori algorithm, we can learn the grocery items that are purchased together a.k.a association rules.

The power of this is:

You can find those items that tend to be purchased together more frequently than other items—the ultimate goal being to get shoppers to buy more. Together, these items are called itemsets.

For example:

You can probably quickly see that chips + dip and chips + soda seem to frequently occur together. These are called 2-itemsets. With a large enough dataset, it will be much harder to "see" the relationships especially when you're dealing with 3-itemsets or more. That's precisely what Apriori helps with!

You might be wondering how Apriori works? Before getting into the nitty-gritty of algorithm, you'll need to define 3 things:

(1) The first is the size of your itemset. Do you want to see patterns for a 2-itemset, 3-itemset, etc.?

(2) The second is your support or the number of transactions containing the itemset divided by the total number of transactions. An itemset that meets the support is called a frequent itemset.

(3) The third is your confidence or the conditional probability of some item given you have certain other items in your itemset. A good example is given chips in your itemset, there is a 67% confidence of having soda also in the itemset.

The basic Apriori algorithm is a 3 step approach:

(1) Join. Scan the whole database for how frequent 1-itemsets are.

(2) Prune. Those itemsets that satisfy the support and confidence move onto the next round for 2-itemsets.

(3) Repeat. This is repeated for each itemset level until we reach our previously defined size.

Is this supervised or unsupervised? Apriori is generally considered an unsupervised learning approach, since it's often used to discover or mine for interesting patterns and relationships.

But wait, there's more…

Apriori can also be modified to do classification based on labelled data.

Why use Apriori? Apriori is well understood, easy to implement and has many derivatives.

On the other hand…

The algorithm can be quite memory, space and time intensive when generating itemsets.

Where is it used? Plenty of implementations of Apriori are available. Some popular ones are the ARtool, Weka, and Orange.

5. EM

What does it do? In data mining, expectation-maximization (EM) is generally used as a clustering algorithm (like k-means) for knowledge discovery.

In statistics, the EM algorithm iterates and optimizes the likelihood of seeing observed data while estimating the parameters of a statistical model with unobserved variables.

OK, hang on while I explain…

I'm not a statistician, so hopefully my simplification is both correct and helps with understanding.

Here are a few concepts that will make this way easier…

What's a statistical model? I see a model as something that describes how observed data is generated. For example, the grades for an exam could fit a bell curve, so the assumption that the grades are generated via a bell curve (a.k.a. normal distribution) is the model.

Wait, what's a distribution? A distribution represents the probabilities for all measurable outcomes. For example, the grades for an exam could fit a normal distribution. This normal distribution represents all the probabilities of a grade.

In other words, given a grade, you can use the distribution to determine how many exam takers are expected to get that grade.

Cool, what are the parameters of a model? A parameter describes a distribution which is part of a model. For example, a bell curve can be described by its mean and variance.

Using the exam scenario, the distribution of grades on an exam (the measurable outcomes) followed a bell curve (this is the distribution). The mean was 85 and the variance was 100.

So, all you need to describe a normal distribution are 2 parameters: The mean, The variance.

And likelihood? Going back to our previous bell curve example… suppose we have a bunch of grades and are told the grades follow a bell curve. However, we're not given all the grades… only a sample.

Here's the deal:

We don't know the mean or variance of all the grades, but we can estimate them using the sample. The likelihood is the probability that the bell curve with estimated mean and variance results in those bunch of grades.

In other words, given a set of measurable outcomes, let's estimate the parameters. Using these estimated parameters, the hypothetical probability of the outcomes is called likelihood.

Remember, it's the hypothetical probability of the existing grades, not the probability of a future grade.

You're probably wondering, what's probability then?

Using the bell curve example, suppose we know the mean and variance. Then we're told the grades follow a bell curve. The chance that we observe certain grades and how often they are observed is the probability.

In more general terms, given the parameters, let's estimate what outcomes should be observed. That's what probability does for us.

Great! Now, what's the difference between observed and unobserved data? Observed data is the data that you saw or recorded. Unobserved data is data that is missing. There a number of reasons that the data could be missing (not recorded, ignored, etc.).

Here's the kicker:

For data mining and clustering, what's important to us is looking at the class of a data point as missing data. We don't know the class, so interpreting missing data this way is crucial for applying EM to the task of clustering.

Once again: The EM algorithm iterates and optimizes the likelihood of seeing observed data while estimating the parameters of a statistical model with unobserved variables. Hopefully, this is way more understandable now.

The best part is…

By optimizing the likelihood, EM generates an awesome model that assigns class labels to data points—sounds like clustering to me!

How does EM help with clustering? EM begins by making a guess at the model parameters.

Then it follows an iterative 3-step process:

(1) E-step: Based on the model parameters, it calculates the probabilities for assignments of each data point to a cluster.

(2) M-step: Update the model parameters based on the cluster assignments from the E-step.

(3) Repeat until the model parameters and cluster assignments stabilize (a.k.a. convergence).

Is this supervised or unsupervised? Since we do not provide labeled class information, this is unsupervised learning.

Why use EM? A key selling point of EM is it's simple and straight-forward to implement. In addition, not only can it optimize for model parameters, it can also iteratively make guesses about missing data.

This makes it great for clustering and generating a model with parameters. Knowing the clusters and model parameters, it's possible to reason about what the clusters have in common and which cluster new data belongs to.

EM is not without weaknesses though…

(1) First, EM is fast in the early iterations, but slow in the later iterations.

(2) Second, EM doesn't always find the optimal parameters and gets stuck in local optima rather than global optima.

Where is it used? The EM algorithm is available in Weka. R has an implementation in the mclust package. Scikit-learn also has an implementation in its gmm module.

6. PageRank

PageRank is a link analysis algorithm designed to determine the relative importance of some object linked within a network of objects.

Yikes. What's link analysis? It's a type of network analysis looking to explore the associations (a.k.a. links) among objects.

Here's an example: The most prevalent example of PageRank is Google's search engine. Although their search engine doesn't solely rely on PageRank, it's one of the measures Google uses to determine a web page's importance.

Let me explain:

Web pages on the World Wide Web link to each other. If rayli.net links to a web page on CNN, a vote is added for the CNN page indicating rayli.net finds the CNN web page relevant.

And it doesn't stop there…

rayli.net's votes are in turn weighted by rayli.net's importance and relevance. In other words, any web page that's voted for rayli.net increases rayli.net's relevance.

The bottom line?

This concept of voting and relevance is PageRank. rayli.net's vote for CNN increases CNN's PageRank, and the strength of rayli.net's PageRank influences how much its vote affects CNN's PageRank.

What does a PageRank of 0, 1, 2, 3, etc. mean? Although the precise meaning of a PageRank number isn't disclosed by Google, we can get a sense of its relative meaning.

You see?

It's a bit like a popularity contest. We all have a sense of which websites are relevant and popular in our minds. PageRank is just an elegant way to define it.

What other applications are there of PageRank? PageRank was specifically designed for the World Wide Web.

Think about it:

At its core, PageRank is really just a super effective way to do link analysis. The objects being linked don't have to be web pages.

Is this supervised or unsupervised? PageRank is generally considered an unsupervised learning approach, since it's often used to discover the importance or relevance of a web page.

Why use PageRank? Arguably, the main selling point of PageRank is its robustness due to the difficulty of getting a relevant incoming link.

Simply stated:

If you have a graph or network and want to understand relative importance, priority, ranking or relevance, give PageRank a try.

Where is it used? The PageRank trademark is owned by Google. However, the PageRank algorithm is actually patented by Stanford University.

New Words

classifier	[ˈklæsifaiə]	n.分类者；分类器
deal	[diːl]	vi.处理
flowchart	[ˈfləutʃɑːt]	n.流程图，程序框图
supervise	[ˈsjuːpəvaiz]	v.监督，管理，指导
unsupervise	[ˌʌnˈsjuːpəvaiz]	v.无监督，不管理，不指导
generate	[ˈdʒenəˌreit]	vt.产生，发生
pruning	[ˈpruːniŋ]	n.修枝，剪枝，修剪
over-fitting	[ˈəuvə-ˈfitiŋ]	n.过拟合
readable	[ˈriːdəbl]	adj.易读的
checkout	[ˈtʃekaut]	v.检验，校验
cluster	[ˈklʌstə]	n.簇，串，丛 vi.丛生，成群
synonymous	[siˈnɔniməs]	adj.同义的
observation	[ˌɔbzəːˈveiʃən]	n.观察资料（或报告）；观察，观测
optimize	[ˈɔptimaiz]	vt.使最优化
centroid	[ˈsentrɔid]	n.质心，矩心
stabilize	[ˈsteibilaiz]	v.稳定
convergence	[kənˈvəːdʒəns]	n.收敛，集中

simplicity	[sim'plisiti]	n.简单，简易
massive	['mæsiv]	adj.大量的，厚重的，大块的
sensitivity	['sensi'tiviti]	n.敏感；灵敏度，灵敏性
trick	[trik]	n.窍门，诀窍
impress	[im'pres]	vt.留下印象
hyperplane	['haipəplein]	n.超平面
complicated	['kɔmplikeitid]	adj.复杂的，难解的
kernel	['kə:nl]	n.内核，核心，要点，精髓
misclassification	[ˌmisklæsifi'keiʃən]	n.错分类，误分类
equidistant	[ˌi:kwi'distənt]	adj.距离相等的，等距的
itemset	['aitemset]	n.项目集
nitty-gritty	['niti'griti]	n.事实真相，本质
prune	[pru:n]	v.剪除
satisfy	['sætisfai]	v.满意，确保
derivative	[di'rivətiv]	adj.引出的，导出的
		n.派生的事物；派生词
estimate	['estimeit]	v. & n.估计，估价，评估
unobserved	[ˌʌnəb'zə:vd]	adj.没有观察到的，无人遵守的
statistician	[ˌstæti'stiʃən]	n.统计员，统计学家
simplification	[ˌsimplifi'keiʃən]	n.简化
measurable	['meʒərəbl]	adj.可测量的
distribution	[ˌdistri'bju:ʃən]	n.分布
describe	[dis'kraib]	v.描述
likelihood	['laiklihud]	n.可能，可能性
bunch	[bʌntʃ]	n.串，束
		v.捆成一束
sample	['sæmpl]	n.标本，样品
		vt.取样，采样
hypothetical	[ˌhaipəu'θetikəl]	adj.假设的，假定的
mean	[mi:n]	n.平均数，中间
awesome	['ɔ:səm]	adj.引起敬畏的，可怕的
disclose	[dis'kləuz]	vt.透露
elegant	['eligənt]	adj.第一流的
relevance	['reləvəns]	n.相关性，关联
robustness	[rə'bʌstnəs]	n.鲁棒性，健壮性
graph	[grɑ:f]	n.图表，曲线图
ranking	['ræŋkiŋ]	n.排名

patent	['pætənt]	n. 专利权，执照，专利品 adj. 特许的，专利的 vt. 取得……的专利权

🍂 Phrases

in the form of…	以……的形式
blood pressure	血压
correlate with…	使……与……发生关系，把……与……联系起来
bottom line	底线
information gain	信息增益
single-pass pruning process	单程修剪过程
hang on	坚持；等候
multi-dimensional space	多维空间
supervised learning	监督学习
No Free Lunch Theorem	没有免费的午餐
learning correlation	学习相关
association rules	关联规则
frequent itemset	频繁项集，频繁集，频集
conditional probability	条件概率
labelled data	标记数据
clustering algorithm	类聚算法
statistical model	统计模式
bell curve	钟形曲线，正态分布
normal distribution	正态分布
observed data	观测数据
selling point	卖点
get stuck in	起劲儿地做某事，使劲儿干
link analysis algorithm	链接分析算法
search engine	搜索引擎
World Wide Web	万维网

🍂 Abbreviations

SVM (Support Vector Machine)	支持向量机
a.k.a. (also known as)	又名，也叫作，换句话说

EM (Expectation-Maximization)　　　　最大期望算法

Exercises

【Ex. 5】 根据课文内容回答问题。
1. How does C4.5 construct a classifier?
2. What is a classifier?
3. What does k-means do?
4. What's cluster analysis?
5. What is the key selling point of k-means? What does it mean?
6. What does SVM do at a high-level?
7. What is the margin?
8. What are a few of the popular implementations of SVM?
9. What does the Apriori algorithm do?
10. Why is Apriori generally considered an unsupervised learning approach?
11. What is expectation-maximization (EM) generally used as in data mining?
12. What does the EM algorithm do in statistics?
13. What's the difference between observed and unobserved data?
14. What is PageRank?
15. What's link analysis?

参考译文

数 据 挖 掘

数据挖掘是一个功能强大的新技术，它具有巨大潜力，可以帮助企业专注于其所收集的客户和潜在客户的行为数据中最重要的信息。它能够发现查询和报表数据中不能有效揭示的信息。

1. 什么是数据挖掘

数据挖掘或知识发现是一种计算机辅助方法，它挖掘和分析巨量的数据集然后提取数据的意义。数据挖掘工具可以预测行为和未来的发展趋势，使企业做出积极主动的知识驱动的决策。数据挖掘工具可以回答传统上要耗费大量时间才能解决的业务问题。它们搜索数据库中的隐藏模式，寻找专家没有想到而可能会错过的预测信息。

数据挖掘技术从大型数据库中搜索有价值的信息,这与从山脉挖掘宝贵的矿石相似,也因此得名。这两个过程都需要对巨大数量的材料进行筛选,或智能探测其价值所在。

2. 挖掘可以做什么

尽管数据挖掘尚处于起步阶段,但已经使用数据挖掘工具和技术的公司却广泛于各个行业——包括零售、金融、卫生保健、制造运输和航空航天——它们都已经使用数据挖掘工具利用历史数据。通过使用模式识别技术以及统计和数学方法来筛选信息仓库,数据挖掘帮助分析师识别重要的事实、关系、趋势、模式以及可能会被忽视的例外和异常情况。

对于企业来说,数据挖掘是用来发现数据中的模式和关系以帮助做出更好的业务决策。数据挖掘技术可以帮助发现销售趋势,制订更明智的营销活动计划,并准确地预测客户的忠诚度。数据挖掘的具体用途包括:

- 市场细分——识别从公司购买相同产品的客户的共同特点。
- 客户流失——预测哪些客户有可能离开公司去购买竞争对手的产品。
- 欺诈检测——确定哪些交易是最有可能是欺诈。
- 直销——确定应包含在邮件列表中的产品,以获得最高的响应速度。
- 互动营销——预测每个人访问网站时可能最感兴趣的内容。
- 市场购物篮分析——了解什么样的产品或服务通常一起购买;例如,啤酒和尿布。
- 趋势分析——显示一个典型的客户本月与上月的不同。

数据挖掘技术可以通过以下方式创造新的商业机会:

自动预测趋势和行为——数据挖掘在一个大的数据库自动发现预测信息。传统上需要大量人工分析的问题,现在可以直接从数据中得到答案。预测问题的典型例子是目标营销。数据挖掘使用过去的促销邮件数据,以确定将来邮件中最可能获得最大回报的目标人群。其他预测问题包括预测破产和其他默认形式,和可能对特定事件做出相同回应的人员范围。

自动发现以前未知的模式——数据挖掘工具扫描数据库并确定以前未见的模式。模式发现的一个例子是分析零售数据,找出那些经常一起购买的看似无关的产品。其他模式发现的问题包括检测欺诈性信用卡交易并识别录入错误所产生的异常数据。

采用大规模并行计算机,企业通过挖掘大量数据,发现他们的客户和产品模式。例如,杂货连锁店已经发现,当男人去超市买尿布,他们有时也带走一包六瓶的啤酒。利用这些信息,就可能重新摆放货物,让这些商品的位置更近。

AT&T 公司、AC 尼尔森和美国运通公司正在销售和营销中率先应用数据挖掘技术。这些系统通过对吉字节的销售点数据运算,来帮助分析师了解消费者行为和制订促销策略。为什么呢?为了获得竞争优势,提高盈利能力!

同样，金融分析师通过对大量的财务记录、流入数据和其他信息源进行分析，以做出投资决定。医疗机构正在审查医疗记录，以了解过去的趋势，以便在未来降低成本。

3. 数据挖掘如何工作

数据挖掘为何能够告诉你不知道的事情或者接下来会发生的事情？那是因为它使用了称为建模的技术。建模就是简单地基于已知情况的数据建立模型（一组例子或数学关系），然后将模型应用到未知答案的其他情况中。建模技术已经存在了几个世纪，当然，只是在最近才具有了数据存储以及收集和存储大量数据所需的通信能力，并能提供自动建模技术直接使用数据所需的计算能力。

假定电信公司的营销总监要构建模型。他想把营销和销售集中于最有可能成为长途电话的大用户的人群。他对客户了解不少，但不能辨别最好客户的共同特点，因为变化因素众多。他可以使用数据挖掘工具（如神经网络）从现有客户数据库（其中包含如年龄、性别、信用记录、收入、邮编、职业等信息）来确定大量的长途电话客户的特点。例如，他可能知道他最好的客户是34～42岁的未婚女性，每年话费超过60000美元。那么，这就是高价值客户模型，他将据此调整自己的营销预算。

4. 数据挖掘技术

在数据挖掘中所用的分析技术就是众所周知的数学算法和技术。它的新颖之处是可以通过增加的数据使用和廉价存储以及处理能力的提高，把这些技术应用到解决一般业务问题上。

此外，使用图形界面使得业务专家也能轻松使用工具。

一些用于数据挖掘的工具有：

人工神经网络——学习和模拟生物神经网络结构的非线性预测模型。

决策树——表示决策集的树形结构。它们生成用于数据集分类的决策规则。

规则归纳——从基于统计学意义的数据中提取有用的"如果-那么"规则。

遗传算法——基于遗传组合、变异和自然选择的概念优化技术。

最邻近算法——一种分级技术，把历史数据库中的每个记录按照相似性进行分类。

5. 真实世界的例子

对于服务目标客户和设备的特定客户来说，谁呼叫谁、他们通话多长时间以及线路是否被用于传真以及语音这些细节可以是无价的。但这些细节都埋藏在数据库中的众多数字之中。通过深入研究广泛的客户呼叫数据库来管理其通信网络，区域电话公司可以识别新型的未满足需求的客户。利用其数据挖掘系统，它发现了如何通过测量家庭在选定时间内日常使用电话的情况来确定可能的附加服务。例如，在下午3点和下午6点之间有许多长时间的通话的那些家庭中，可能主要是家里的青少年在打电话。当公司使用目标市场营销，强调成年人的便利性和价值时——"手机总是占线？"——隐藏的需求就显示出来了。上午9点到下午5点电话使用的特点是使用语音、传真模式和调制解调器，这表明客户在进行业务活动。目标市场销售要为这些客户提供"小预算业务通信性能"的服务，就要增加附加线、功能和设备的销售。

能够准确地衡量客户对业务规则变化的响应能力是一个强大的竞争优势。银行根据信用卡操作来寻找增加收入的新方法，可以测试其非直观性：如果银行将其最低要求付款减半会使信用卡和利息收入显著增加吗？由于两年期间平均信用卡余额、付款金额、付款及时性、信用额度的使用情况以及其他关键参数共计数百吉字节数据，该行采用了功能强大的数据挖掘系统来模拟所提出的政策变化对特定客户的影响。该银行发现，如果取消最低付款要求，对小公司和目标客户可能会增加平均余额和延长债务期限，产生超过 2500 万美元的额外利息。Merck-Medco Managed Care（默德克-梅德科管理保健公司）是一家邮购公司，其主要业务是向全国最大的医疗保健提供者销售药品。Medco 公司正在挖掘其一太字节（TB）的数据仓库以便发现疾病和已知的药物治疗之间隐藏的关联，帮助确定哪些药物对什么类型的患者是最有效的。结果是，更有效的治疗也更便宜。Merck-Medco 的数据挖掘项目已经帮助客户节省了平均 10%～15% 的药费。

6. 数据挖掘的未来

在短期内，数据挖掘的结果将用于盈利性的业务领域。微营销活动将探索新的商机。广告将更精准地瞄准潜在客户。

从中期来看，数据挖掘可以如电子邮件一样普通和易用。我们可以使用这些工具来寻找到纽约的最佳机票，深挖了久违的同学的电话号码，或找到最好价格的割草机。

从长期来看，前景是真正令人兴奋的。想象一下，智能代理放开了医学研究资料或亚原子粒子数据。计算机可以揭示疾病治疗的新方法。

Unit 10

Text A

What Is Hadoop?

Everyone's talking about Hadoop, the hot new technology that's highly prized among developers and just might change the world (again). But just what is it, anyway? Is it a programming language? A database? A processing system? An Indian tea cozy?

The broad answer: Hadoop is all of these things (except the tea cozy), and more. It's a software library that provides a programming framework for cheap, useful processing of another modern buzzword: big data.

1. Where did Hadoop come from?

Apache Hadoop is part of the Foundation Project from the Apache Software Foundation, a non-profit organization whose mission is to "provide software for the public good." As such, the Hadoop library is a free, open-source software available to all developers.

The underlying technology that powers Hadoop was actually invented by Google. Back in the early days, the not-quite-giant search engine needed a way to index the massive amounts of data they were collecting from the Internet, and turn it into meaningful, relevant results for its users. With nothing available on the market that could meet their requirements, Google built their own platform.

Those innovations were released in an open-source project called Nutch, which Hadoop later used as a foundation. Essentially, Hadoop applies the power of Google to big data in a way that's affordable for companies of all sizes.

2. How is Hadoop Different from Past Techniques?

Hadoop is more than just a faster, cheaper database and analytics tool. Unlike databases, Hadoop doesn't insist that you structure your data. Data may be unstructured and schemaless. Users can dump their data into the framework without needing to reformat it. By contrast, relational databases require that data be structured and schemas be defined before storing the data.

Hadoop's simplified programming model allows users to quickly write and test software in distributed systems. Performing computation on large volumes of data has been done before, usually in a distributed setting but writing software for distributed systems is notoriously hard. By trading away some programming flexibility, Hadoop makes it much easier to write distributed programs.

Because Hadoop accepts practically any kind of data, it stores information in far more diverse formats than what is typically found in the tidy rows and columns of a traditional database. Some good examples are machine-generated data and log data, written out in storage formats including JSON, Avro and ORC.

The majority of data preparation work in Hadoop is currently being done by writing code in scripting languages like Hive, Pig or Python.

Hadoop is easy to administer.

Alternative high performance computing (HPC) systems allow programs to run on large collections of computers, but they typically require rigid program configuration and generally require that data be stored on a separate storage area network (SAN) system. Schedulers on HPC clusters require careful administration and since program execution is sensitive to node failure, administration of a Hadoop cluster is much easier.

Hadoop invisibly handles job control issues such as node failure. If a node fails, Hadoop makes sure the computations are run on other nodes and that data stored on that node are recovered from other nodes.

Hadoop is agile.

Relational databases are good at storing and processing data sets with predefined and rigid data models. For unstructured data, relational databases lack the agility and scalability that are needed. Apache Hadoop makes it possible to cheaply process and analyze huge amounts of both structured and unstructured data together, and to process data without defining all structure ahead of time.

3. Why use Apache Hadoop?

Apache Hadoop controls costs by storing data more affordably per terabyte than other platforms. Instead of thousands to tens of thousands per terabyte, Hadoop delivers compute and storage for hundreds of dollars per terabyte.

Fault tolerance is one of the most important advantages of using Hadoop. Even if individual nodes experience high rates of failure when running jobs on a large cluster, data is replicated across a cluster so that it can be recovered easily in the face of disk, node or rack failures.

It's flexible.

The flexible way that data is stored in Apache Hadoop is one of its biggest assets — enabling businesses to generate value from data that was previously considered too expensive to be stored and processed in traditional databases. With Hadoop, you can use all types of data, both structured and unstructured, to extract more meaningful business insights from more of your data.

It's scalable.

Hadoop is a highly scalable storage platform, because it can store and distribute very large data sets across clusters of hundreds of inexpensive servers operating in parallel. The problem with traditional relational database management systems (RDBMS) is that they can't scale to process massive volumes of data.

4. How does Hadoop work?

As mentioned previously, Hadoop isn't one thing—it's many things. Hadoop is a software library, which consists of four primary parts (modules), and a number of add-on solutions (like databases and programming languages) that enhance its real-world use. The four modules are:

- Hadoop Common: This is the collection of common utilities (the common library) that supports Hadoop modules.
- Hadoop Distributed File System (HDFS): A robust distributed file system with no restrictions on stored data (meaning that data can be either structured or unstructured and schemaless, where many DFSs will only store structured data) that provides high-throughput access with redundancy (HDFS allows data to be stored on multiple machines — so if one machine fails, availability is maintained through the other machines).

- Hadoop YARN: This framework is responsible for job scheduling and cluster resource management; it makes sure the data is spread out sufficiently over multiple machines to maintain redundancy. YARN is the module that makes Hadoop an affordable and cost-efficient way to process big data.
- Hadoop MapReduce: This YARN-based system, built on Google technology, carries out parallel processing of large data sets (structured and unstructured). MapReduce can also be found in most of today's big data processing frameworks, including MPP and NoSQL databases.

All of these modules working together generate distributed processing for large data sets. The Hadoop framework uses simple programming models that are replicated across clusters of computers, meaning the system can scale up from single servers to thousands of machines for increased processing power, rather than relying on hardware alone.

Hardware that can handle the amount of processing power required to work with big data is expensive, to put it mildly. This is the true innovation of Hadoop: the ability to break down massive amounts of processing power across multiple, smaller machines, each with its own localized computation and storage, along with built-in redundancy at the application level to prevent failures.

New Words

developer	[di'veləpə]	n.开发者
buzzword	['bʌzwə:d]	n.时髦术语，流行行话，新潮词汇
mission	['miʃən]	n.使命，任务
meaningful	['mi:niŋful]	adj.有意义的，有意图的；意味深长的
platform	['plætfɔ:m]	n.平台
essentially	[i'senʃəli]	adv.本质上，本来
insist	[in'sist]	v.坚持，强调
structure	['strʌktʃə]	vt.构成，组织
		n.结构，构造
dump	[dʌmp]	vt.转存；倾倒
		n.堆存处
notorious	[nəu'tɔ:riəs]	adj.臭名远扬的；恶名昭著的
diverse	[dai'və:s]	adj.不同的，变化多的
rigid	['ridʒid]	adj.严格的
scheduler	['ʃedju:lə]	n.调度程序
agile	['ædʒail]	adj.敏捷的，轻快的，灵活的
predefine	['pri:di'fain]	vt.预先确定，预先定义

agility	[əˈdʒiliti]	n.敏捷性
scalability	[ˌskeiləˈbiliti]	n.可伸展性
Terabyte	[ˈterəbait]	n.太字节（TB）。1TB=1024GB=2^{40}B
replicate	[ˈreplikeit]	v.复制
scalable	[ˈskeiləbl]	adj.可升级的
utility	[juːˈtiliti]	n.效用，有用，实用
restriction	[risˈtrikʃən]	n.限制，约束
throughput	[ˈθruːput]	n.吞吐量，生产量，生产能力
availability	[əˌveiləˈbiliti]	n.可用性，有效性，实用性
localize	[ˈləukəlaiz]	v.（使）局部化，本地化

Phrases

programming language	编程语言
tea cozy	茶壶套
software library	软件库
Apache Software Foundation	Apache 软件基金会（简称为 ASF）
non-profit organization	非盈利组织
free，open-source software	免费开源软件
distributed system	分布式系统
machine-generated data	机器生成的数据
log data	日志数据
data preparation	数据准备
scripting language	脚本语言
node failure	节点失效，节点故障；点失效
fault tolerance	容错
job scheduling	作业调度
cluster resource management	集群资源管理
spread out	分散，展开
carry out	完成，实现，执行
parallel processing	并行处理
scale up	按比例增加，按比例提高
put it mildly	说得委婉些，说得好听一点

Abbreviations

HPC (High Performance Computing)	高性能计算
SAN (Storage Area Network)	存储区域网络

RDBMS (Relational DataBase Management System)　　关系型数据库管理系统
HDFS (Hadoop Distributed File System)　　Hadoop 分布式文件处理系统
DFS (Distributed File System)　　分布式文件处理系统
MPP (Massive Parallel Processing)　　大规模并行处理

Notes

[1] Everyone's talking about Hadoop, the hot new technology that's highly prized among developers and just might change the world (again).

本句中，the hot new technology that's highly prized among developers and just might change the world (again)是一个名词性短语，对 Hadoop 进行补充说明。在该从句中，that's highly prized among developers and just might change the world (again)是一个定语从句，修饰和限定 the hot new technology。

[2] Those innovations were released in an open-source project called Nutch, which Hadoop later used as a foundation.

本句中，called Nutch 是一个过去分词短语，作定语，修饰和限定 an open-source project。which Hadoop later used as a foundation 是一个非限定性定语从句，对 Those innovations 进行补充说明。

[3] Even if individual nodes experience high rates of failure when running jobs on a large cluster, data is replicated across a cluster so that it can be recovered easily in the face of disk, node or rack failures.

本句中，Even if individual nodes experience high rates of failure when running jobs on a large cluster 是一个让步状语从句，修饰主句的谓语 is replicated。在该从句中，when running jobs on a large cluster 是一个时间状语从句，修饰从句的谓语 experience。so that it can be recovered easily in the face of disk, node or rack failures 是一个目的状语从句，修饰主句的谓语 is replicated。

[4] Hadoop is a software library, which consists of four primary parts (modules), and a number of add-on solutions (like databases and programming languages) that enhance its real-world use.

本句中，which consists of four primary parts (modules), and a number of add-on solutions (like databases and programming languages) that enhance its real-world use 是一个非限定性定语从句，对 a software library 进行补充说明。that enhance its real-world use 是一个定语从句，修饰和限定 four primary parts (modules), and a number of add-on solutions。(like databases and programming languages)对 add-on solutions 举例说明。

[5] The Hadoop framework uses simple programming models that are replicated across clusters of computers, meaning the system can scale up from single servers to thousands of machines for increased processing power, rather than relying on hardware alone.

本句中，that are replicated across clusters of computers 是一个定语从句，修饰和限定 simple programming models。meaning the system can scale up from single servers to thousands of machines for increased processing power, rather than relying on hardware alone 是对前面整个句子的解释说明，可以扩展为一个非限定性定语从句：which means the system can scale up from single servers to thousands of machines for increased processing power, rather than relying on hardware alone.

Exercises

【Ex. 1】根据课文内容回答问题。

1. What is Apache Hadoop?
2. What did the not-quite-giant search engine need back in the early days?
3. What does Hadoop's simplified programming model allow users to do?
4. Why does Hadoop store information in far more diverse formats than what is typically found in the tidy rows and columns of a traditional database?
5. What does Hadoop do if a node fails?
6. What is one of the most important advantages of using Hadoop?
7. Why is Hadoop is a highly scalable storage platform?
8. What is the problem with traditional relational database management systems (RDBMS)?
9. What are the four primary modules that Hadoop consists of?
10. What is the true innovation of Hadoop?

【Ex. 2】把下列句子翻译为中文。

1. Using this method, each developer can provide their own physical path definition to this variable.
2. All the data is then dumped into the main computer.
3. All of the configuration and code is already implemented in the sample.
4. This is about the simplest weightless thread scheduler you could choose.
5. Such models align with agile thinking.
6. This can result in a variety of scalability and maintenance problems.
7. This allows the storage nodes to replicate data when a device is found to have failed.
8. Scalable bandwidth provides the solution while offering a more efficient use of network resources.
9. Redundancy and dependability give the cloud another edge.
10. The most fundamental reason for a software company to localize product is to increase total revenue and net income.

【Ex. 3】 短文翻译。

1. What are the challenges of using Hadoop?

MapReduce programming is not a good match for all problems. It's good for simple information requests and problems that can be divided into independent units, but it's not efficient for iterative and interactive analytic tasks. MapReduce is file-intensive. Because the nodes don't intercommunicate except through sorts and shuffles, iterative algorithms require multiple map-shuffle/sort-reduce phases to complete. This creates multiple files between MapReduce phases and is inefficient for advanced analytic computing.

There's a widely acknowledged talent gap. It can be difficult to find entry-level programmers who have sufficient Java skills to be productive with MapReduce. That's one reason distribution providers are racing to put relational (SQL) technology on top of Hadoop. It is much easier to find programmers with SQL skills than MapReduce skills. And, Hadoop administration seems part art and part science, requiring low-level knowledge of operating systems, hardware and Hadoop kernel settings.

Data security. Another challenge centers around the fragmented data security issues, though new tools and technologies are surfacing. The Kerberos authentication protocol is a great step toward making Hadoop environments secure.

Full-fledged data management and governance. Hadoop does not have easy-to-use, full-feature tools for data management, data cleansing, governance and metadata. Especially lacking are tools for data quality and standardization.

2. Why is Hadoop important?

Ability to store and process huge amounts of any kind of data, quickly. With data volumes and varieties constantly increasing, especially from social media and the Internet of Things (IoT), that's a key consideration.

Computing power. Hadoop's distributed computing model processes big data fast. The more computing nodes you use, the more processing power you have.

Fault tolerance. Data and application processing are protected against hardware failure. If a node goes down, jobs are automatically redirected to other nodes to make sure the distributed computing does not fail. Multiple copies of all data are stored automatically.

Flexibility. Unlike traditional relational databases, you don't have to preprocess data before storing it. You can store as much data as you want and decide how to use it later. That includes unstructured data like text, images and videos.

Low cost. The open-source framework is free and uses commodity hardware to store large quantities of data.

Scalability. You can easily grow your system to handle more data simply by adding nodes. Little administration is required.

【Ex. 4】将下列词填入适当的位置（每词只用一次）。

| bottom | duplicate | machines | special | node |
| source | collected | operations | completion | individual |

1. Background of Hadoop

With an increase in the penetration of internet and the usage of the internet, the data captured by Google increased exponentially year on year. Just to give you an estimate of this number, in 2007 Google __(1)__ on an average 270 PB of data every month. The same number increased to 20000 PB everyday in 2009. Obviously, Google needed a better platform to process such an enormous data. Google implemented a programming model called MapReduce, which could process this 20000 PB per day. Google ran these MapReduce operations on a __(2)__ special file system called Google File System (GFS). Sadly, GFS is not an open source.

Doug cutting and Yahoo! reverse engineered the model GFS and built a parallel Hadoop Distributed File System (HDFS). The software or framework that supports HDFS and MapReduce is known as Hadoop. Hadoop is an open __(3)__ and distributed by Apache.

2. Framework of Hadoop Processing

Let's draw an analogy from our daily life to understand the working of Hadoop. The bottom of the pyramid of any firm are the people who are __(4)__ contributors. They can be analyst, programmers, manual labors, chefs, etc. Managing their work is the project manager. The project manager is responsible for a successful __(5)__ of the task. He needs to distribute labor, smoothen the coordination among them etc. Also, most of these firms have a people manager, who is more concerned about retaining the head count.

Hadoop works in a similar format. On the __(6)__ we have machines arranged in parallel. These machines are analogous to individual contributor in our analogy. Every machine has a data node and a task tracker. Data node is also known as HDFS (Hadoop Distributed File System) and task tracker is also known as map-reducers.

Data node contains the entire set of data and task tracker does all the __(7)__. You can imagine task tracker as your arms and leg, which enables you to do a task and data node as your brain, which contains all the information which you want to process. These __(8)__ are working in silos and it is very essential to coordinate them. The task trackers (project manager in our analogy) in different machines are coordinated by a job tracker. Job tracker makes sure that each operation is completed and if there is a process failure at any __(9)__, it needs to assign a duplicate task to some task tracker. Job tracker also distributes the entire task to all the machines.

A name node on the other hand coordinates all the data nodes. It governs the distribution of data going to each machine. It also checks for any kind of purging which have happened on any machine. If such purging happens, it finds the __(10)__ data which was sent to other data node and duplicates it again. You can think of this name node as the people manager in our analogy which is concerned more about the retention of the entire dataset.

Text B

Apache Spark

Apache Spark is an open source parallel processing framework for running large-scale data analytics applications across clustered computers. It can handle both batch and real-time analytics and data processing workloads.

Spark became a top-level project of the Apache Software Foundation in February 2014, and version 1.0 of Apache Spark was released in May 2014. Spark version 2.0 was released in July 2016.

The technology was initially designed in 2009 by researchers at the University of California, Berkeley as a way to speed up processing jobs in Hadoop systems.

Spark Core, the heart of the project that provides distributed task transmission, scheduling and I/O functionality, provides programmers with a potentially faster and more flexible alternative to MapReduce. MapReduce is the software framework to which early versions of Hadoop were tied. Spark's developers say it can run jobs 100 times faster than MapReduce when processed in memory, and 10 times faster on disk.

1. How Apache Spark works

Apache Spark can process data from a variety of data repositories, including the Hadoop Distributed File System (HDFS), NoSQL databases and relational data stores, such as Apache Hive. Spark supports in-memory processing to boost the performance of big data analytics applications, but it can also perform conventional disk-based processing when data sets are too large to fit into the available system memory.

The Spark Core engine uses the resilient distributed data set, or RDD, as its basic data type. The RDD is designed in such a way so as to hide much of the computational complexity from users. It aggregates data and partitions it across a server cluster, where it can then be computed and either moved to a different data store or run through an analytic model. The user doesn't have to define where specific files are sent or what computational resources are used to store or retrieve files.

In addition, Spark can handle more batch processing applications than MapReduce.

2. Spark libraries

The Spark Core engine functions partly as an application programming interface (API) layer and underpins a set of related tools for managing and analyzing data. Aside from the Spark Core processing engine, the Apache Spark API environment comes packaged with some libraries of code for use in data analytics applications.

2.1 Spark Core

Spark Core is the foundation of the overall project. It provides distributed task dispatching, scheduling, and basic I/O functionalities, exposed through an application programming interface (for Java, Python, Scala, and R) centered on the RDD abstraction (the Java API is available for other JVM languages, but is also usable for some other non-JVM languages, such as Julia, that can connect to the JVM). This interface mirrors a functional/higher-order model of programming: a "driver" program invokes parallel operations such as map, filter or reduce on an RDD by passing a function to Spark, which then schedules the function's execution in parallel on the cluster. These operations, and additional ones such as joins, take RDDs as input and produce new RDDs. RDDs are immutable and their operations are lazy; fault-tolerance is achieved by keeping track of the "lineage" of each RDD (the sequence of operations that produced it) so that it can be reconstructed in the case of data loss. RDDs can contain any type of Python, Java, or Scala objects.

Besides the RDD-oriented functional style of programming, Spark provides two restricted forms of shared variables: broadcast variables and accumulators. Broadcast variables reference read-only data that needs to be available on all nodes, while accumulators can be used to program reductions in an imperative style. Transform an RDD into a new RDD.

2.2 Spark SQL

Spark SQL is a component on top of Spark Core that introduces a data abstraction called

DataFrames, which provides support for structured and semi-structured data. Spark SQL provides a domain-specific language (DSL) to manipulate DataFrames in Scala, Java, or Python. It also provides SQL language support, with command-line interfaces and ODBC/JDBC server. Although DataFrames lacks the compile-time type-checking afforded by RDDs, as of Spark 2.0, the strongly typed DataSet is fully supported by Spark SQL as well.

2.3 Spark Streaming

Spark Streaming uses Spark Core's fast scheduling capability to perform streaming analytics. It ingests data in mini-batches and performs RDD transformations on those mini-batches of data. This design enables the same set of application code written for batch analytics to be used in streaming analytics, thus facilitating easy implementation of lambda architecture. However, this convenience comes with the penalty of latency equal to the mini-batch duration. Other streaming data engines that process event by event rather than in mini-batches include Storm and the streaming component of Flink. Spark Streaming has support built-in to consume from Kafka, Flume, Twitter, ZeroMQ, Kinesis, and TCP/IP sockets.

In Spark 2.x, a separate technology based on Datasets, called Structured Streaming, that has a higher-level interface is also provided to support streaming.

2.4 MLlib (Machine Learning Library)

Spark MLlib is a distributed machine learning framework on top of Spark Core that, due in large part to the distributed memory-based Spark architecture, is as much as nine times as fast as the disk-based implementation used by Apache Mahout (according to benchmarks done by the MLlib developers against the alternating least squares (ALS) implementations, and before Mahout itself gained a Spark interface), and scales better than Vowpal Wabbit. Many common machine learning and statistical algorithms have been implemented and are shipped with MLlib which simplifies large scale machine learning pipelines, including:
- summary statistics, correlations, stratified sampling, hypothesis testing, random data generation
- classification and regression: support vector machines, logistic regression, linear regression, decision trees, naive Bayes classification
- collaborative filtering techniques including alternating least squares (ALS)
- cluster analysis methods including k-means, and latent Dirichlet allocation (LDA)
- dimensionality reduction techniques such as singular value decomposition (SVD), and principal component analysis (PCA)
- feature extraction and transformation functions

- optimization algorithms such as stochastic gradient descent, limited-memory BFGS (L-BFGS).

2.5 GraphX

GraphX is a distributed graph processing framework on top of Apache Spark. Because it is based on RDDs, which are immutable, graphs are immutable and thus GraphX is unsuitable for graphs that need to be updated, let alone in a transactional manner like a graph database. GraphX provides two separate APIs for implementation of massively parallel algorithms (such as PageRank): a Pregel abstraction, and a more general MapReduce style API. Unlike its predecessor Bagel, which was formally deprecated in Spark 1.6, GraphX has full support for property graphs (graphs where properties can be attached to edges and vertices).

GraphX can be viewed as being the Spark in-memory version of Apache Giraph, which utilized Hadoop disk-based MapReduce.

Like Apache Spark, GraphX initially started as a research project at UC Berkeley's AMPLab and Databricks, and was later donated to the Apache Software Foundation and the Spark project.

3. Spark languages

Spark was written in Scala, which is considered the primary language for interacting with the Spark Core engine. Out of the box, Spark also comes with API connectors for using Java and Python. Java is not considered an optimal language for data engineering or data science, so many users rely on Python, which is simpler and more geared toward data analysis.

There is also an R programming package that users can download and run in Spark. This enables users to run the popular desktop data science language on larger distributed data sets in Spark and to use it to build applications that leverage machine learning algorithms.

4. Apache Spark use cases

The wide range of Spark libraries and its ability to compute data from many different types of data stores means Spark can be applied to many different problems in many industries. Digital advertising companies use it to maintain databases of web activity and design campaigns tailored to specific consumers. Financial companies use it to ingest financial data and run models to guide investing activity. Consumer goods companies use it to aggregate customer data and forecast trends to guide inventory decisions and spot new market opportunities.

Large enterprises that work with big data applications use Spark because of its speed and its ability to tie together multiple types of databases and to run different kinds of analytics applications.

New Words

workload	[ˈwəːkləud]	n.工作量
conventional	[kənˈvenʃənl]	adj.常规的，传统的
engine	[ˈendʒin]	n.引擎
resilient	[riˈziliənt]	adj.可恢复的；弹回的
partition	[pɑːˈtiʃən]	n.分割，划分，分开；隔离物
		vt.区分，隔开，分割
retrieve	[riˈtriːv]	v.恢复，重新得到
underpin	[ˌʌndəˈpin]	v.加强……的基础，巩固，支撑
fault-tolerance	[fɔːlt-ˈtɔlərəns]	n.容错
lineage	[ˈliniidʒ]	n.血统，世系
reconstructed	[ˌriːkənˈstrʌktɪd]	adj.重建的，改造的
accumulator	[əˈkjuːmjuleitə]	n.累加器
abstraction	[æbˈstrækʃən]	n.提取
ingest	[inˈdʒest]	vt.摄取，获取，吸收
penalty	[ˈpenəlti]	n.害处
duration	[djuəˈreiʃən]	n.持续时间，为期
pipeline	[ˈpaipˌlain]	n.管道；传递途径
correlation	[ˌkɔriˈleiʃən]	n.相互关系，相关（性）
random	[ˈrændəm]	adj.随机的
regression	[riˈgreʃən]	n.衰退
dimensionality	[diˌmenʃəˈnæliti]	n.维度
stochastic	[stəuˈkæstik]	adj.随机的
unsuitable	[ʌnˈsjuːtəbl]	adj.不适合的，不相称的
predecessor	[ˈpriːdisesə]	n.前辈，前任；（被取代的）原有事物
connector	[kəˈnəktə]	n.连接器
invest	[inˈvest]	v.投资

Phrases

open source	开放源代码，开源
parallel processing	并行处理

fit into	适合
computational complexity	计算的复杂性
batch processing	批处理
parallel operation	并行操作
broadcast variable	广播变量
imperative style	强制方式，命令式风格
command-line interface	命令行界面
lambda architecture	λ结构
distributed machine learning framework	分布式机器学习框架
large scale machine learning pipelines	大规模机器学习流水线，大规模机器学习管道
stratified sample	分层取样
hypothesis testing	假设检验
support vector machine	支持向量机
logistic regression	逻辑回归
linear regression	线形回归
decision tree	决策树，分层次决策
naive Bayes classification	朴素贝叶斯分类
dimensionality reduction techniques	降维技术
optimization algorithm	最优化算法
stochastic gradient descent	随机梯度下降
limited-memory BFGS (L-BFGS)	内存受限的BFGS算法
distributed graph processing framework	分布式图形处理结构
interact with…	与……相合
distributed data set	分布式数据集
consumer goods	生活消费品

Abbreviations

HDFS (Hadoop Distributed File System)	Hadoop分布式文件系统
RDD (Resilient Distributed Datasets)	弹性分布式数据集
API (application programming interface)	应用程序接口
I/O (Input/Output)	输入/输出
JVM (Java Virtual Machine)	Java虚拟机
DSL (Domain-Specific Language)	领域专用语言
ODBC (Open Database Connectivity)	开放数据库连接
JDBC (Java DataBase Connectivity)	Java数据库连接

MLlib (Machine Learning Library)	机器学习库
ALS (alternating least squares)	交替最小二乘
LDA (Latent Dirichlet Allocation)	潜在狄利克雷分布
SVD (Singular Value Decomposition)	奇异值分解
PCA (Principal Component Analysis)	主成分分析
BFGS (Broyden, Fletcher, Goldforb, Shannon)	布罗依丹、弗莱彻、戈德福布、香农四个人名的首字母

Exercises

【Ex. 5】 根据课文内容回答问题。

1. What is Apache Spark?
2. What can Apache Spark do?
3. What does the Spark Core engine use as its basic data type?
4. What is Spark Core?
5. What are the two restricted forms of shared variables Spark provides?
6. What does Spark Streaming use Spark Core's fast scheduling capability to do?
7. What is Spark MLlib?
8. What are the two separate APIs GraphX provides for implementation of massively parallel algorithms?
9. What was Spark written in?
10. What do digital advertising companies and consumer goods companies use Apache Spark to do respectively?

参考译文

什么是 Hadoop

每个人都在谈论 Hadoop，这是开发者非常重视的热门新技术，有可能（再次）改变世界。但是它是什么呢？是编程语言？数据库？一个处理系统？还是印度茶壶套？

宽泛的答案是：Hadoop 是所有这些事情（除了茶壶套），甚至更多。它是一个软件库，提供了一个编程框架，可用来便宜而有用地处理大数据（大数据是另一个现代流行词汇）。

1. Hadoop 来自哪里

Apache Hadoop 是 Apache Software Foundation 基础项目的一部分，该软件基金会是一个非盈利组织，其任务是"为公共事业提供软件"。因此，Hadoop 库是免费的开源软件，可供所有开发人员使用。

Hadoop 的基础技术实际上由谷歌公司研发。早期的时候，搜索引擎并不巨大，它需要一种方式来检索从互联网收集的大量数据，并将其转化为对用户有用的相关结果。谷歌公司在市场上找不到可以满足其需求的产品，就自己建立了平台。

这些创新在一个名为 Nutch 的开放源码项目中发布，后来被用作 Hadoop 的基础。重要的是，Hadoop 将谷歌的强大功能应用于大数据，并提供了适合各种规模公司的方式。

2. Hadoop 与过去的技术有何不同

Hadoop 不仅仅是一个更快、更便宜的数据库和分析工具。与数据库不同，Hadoop 并不强调数据结构。数据可能是非结构化和无模式的。用户可以将其数据转储到框架中，而无须重新格式化。相比之下，关系数据库要求在存储数据之前对数据进行结构化和模式定义。

Hadoop 简化的编程模型允许用户在分布式系统中快速编写和测试软件。以前就可以对大量数据执行计算，但通常要进行分布式设置，要为分布式系统编写软件是非常困难的。通过放弃一些编程灵活性，Hadoop 使编写分布式程序变得更加容易。

由于 Hadoop 几乎可以接受任何类型的数据，它以比传统数据库多得多的格式存储信息，这些数据原来整齐地存储在数据库的行列中。一些很好的例子是机器生成的数据和日志数据，以包含 JSON、Avro 和 ORC 存储格式的数据。

Hadoop 中的大部分数据准备工作目前用脚本语言（如 Hive、Pig 或 Python）编写的程序来完成。

Hadoop 易于管理。

备选的高性能计算（HPC）系统允许程序在大量计算机上运行，但是通常需要严格的程序配置，并且通常要求数据存储在单独的存储区域网络（SAN）系统上。HPC 集群上的调度程序需要精细管理，并且由于程序执行对节点故障十分敏感，所以 Hadoop 集群的管理要容易得多。

Hadoop 默默地处理诸如节点故障之类的作业控制问题。如果节点出现故障，那么 Hadoop 将确保在其他节点上运行计算，并且从其他节点恢复存储在该节点上的数据。

Hadoop 是敏捷的。

关系数据库能很好地存储和处理具有预定义和刚性数据模型的数据集。对于非结构化数据，关系数据库缺乏所需的敏捷性和可扩展性。Apache Hadoop 能够便宜地对大量的结构化和非结构化数据一起处理和分析，并且处理数据时无须提前定义所有结构。

3. 为什么要使用 Apache Hadoop

Apache Hadoop 通过比其他平台更容易地存储每 TB 的数据来控制成本。用 Hadoop 计算和存储每 TB 数据只需花费数百美元，而不用花费数千到数万美元。

容错是 Hadoop 最重要的优点之一。即使单个节点在大型集群上运行作业时遇到很高的故障率，也可以跨集群复制数据，以便在面对磁盘、节点或机架故障时可以轻松恢复。

Hadoop 是灵活的。

灵活地在 Apache Hadoop 中存储数据是其最大的价值，使企业能够从数据中生成价值，而这些数据先前要用昂贵的传统数据库中进行存储和处理。使用 Hadoop，可以使用所有类型的结构化和非结构化数据，因此能够从更多的数据中提取更有意义的业务洞察力。

Hadoop 是可扩展的。

Hadoop 是一个高度可扩展的存储平台，因为它可以在数以百计的并行运行的廉价服务器的群集中存储和分发非常大的数据集。传统关系数据库管理系统（RDBMS）的问题在于它们无法扩展以处理大量数据。

4. Hadoop 如何工作

如前所述，Hadoop 并非只做一件事——而是做很多事情。Hadoop 的软件库由四个主要部分（模块）和许多附加解决方案（如数据库和编程语言）组成，这增强了其实际使用性能。这四个模块是：

Hadoop Common——这是支持 Hadoop 模块的常用工具（通用库）的集合。

- Hadoop 分布式文件系统（HDFS）——一个健壮的分布式文件系统，对存储的数据没有限制（意味着数据可以是结构化的或非结构化的、无模式的，其中许多 DFS 将仅存储结构化数据），其提供了具有冗余的高吞吐量访问（HDFS 允许将数据存储在多台机器上——因此如果一台机器发生故障，则可通过其他机器继续工作）。
- Hadoop YARN——该框架负责作业调度和集群资源管理；它确保数据分散于多台机器以保持冗余。YARN 是 Hadoop 高效经济地处理大数据的模块。
- Hadoop MapReduce——用谷歌技术建立的基于 YARN 的系统，能对大型数据集（结构化和非结构化）进行并行处理。MapReduce 也可以用于当今大多数大型数据处理框架，包括 MPP 和 NoSQL 数据库。

所有这些模块协同工作，对大型数据集进行分布式处理。Hadoop 框架使用在计算机集群中复制的简单编程模型，这意味着系统可以从单个服务器扩展到数千台机器，以提高处理能力，而不是单靠硬件。

能处理大数据的硬件是昂贵的。Hadoop 的真正创新在于：把大量的处理能力分解到多个较小的机器上，每个机器都具有自己的本地化计算和存储能力，同时在应用程序级别内置冗余以防出现故障。

Unit 11

Text A

Data Visualization

Data visualization is viewed by many disciplines as a modern equivalent of visual communication. It involves the creation and study of the visual representation of data, meaning "information that has been abstracted in some schematic form, including attributes or variables for the units of information".

A primary goal of data visualization is to communicate information clearly and efficiently via statistical graphics, plots and information graphics. Numerical data may be encoded using dots, lines, or bars, to visually communicate a quantitative message. Effective visualization helps users analyze and reason about data and evidence. It makes complex data more accessible, understandable and usable. Users may have particular analytical tasks, such as making comparisons or understanding causality, and the design principle of the graphic (i.e., showing comparisons or showing causality) follows the task. Tables are generally used when users will look up a specific measurement, while charts of various types are used to show patterns or relationships in the data for one or more variables.

Data visualization is both an art and a science. It is viewed as a branch of descriptive statistics by some, but also as a grounded theory development tool by others. There is an increasing amount of data created by Internet activity and an expanding number of sensors in the environment. Processing, analyzing and communicating this data present ethical and analytical challenges for data visualization. The data scientists help address this challenge.

1. Overview

Data visualization refers to the techniques used to communicate data or information by encoding it as visual objects (e.g., points, lines or bars) contained in graphics. The goal is to communicate information clearly and efficiently to users. It is one of the steps in data analysis or data science. According to Friedman, the main goal of data visualization is to communicate information clearly and effectively through graphical means. It doesn't mean that data visualization needs to look boring to be functional or extremely sophisticated to look beautiful. To convey ideas effectively, both aesthetic form and functionality need to go hand in hand, providing insights into a rather sparse and complex data set by communicating its key-aspects in a more intuitive way. Yet designers often fail to achieve a balance between form and function, creating gorgeous data visualizations which fail to serve their main purpose — to communicate information.

Indeed, Fernanda Viegas and Martin M. Wattenberg suggested that an ideal visualization should not only communicate clearly, but stimulate viewer engagement and attention.

Data visualization is closely related to information graphics, information visualization, scientific visualization, exploratory data analysis and statistical graphics. In the new millennium, data visualization has become an active area of research, teaching and development (see Figure 11-1).

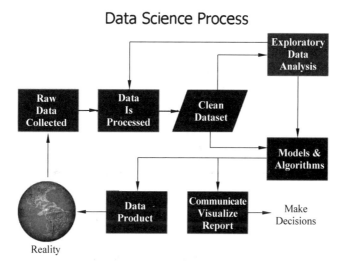

Figure 11-1 Data visualization is one of the steps in analyzing data and presenting it to users.

2. Characteristics of Effective Graphical Displays

Professor Edward Tufte explained that users of information displays are executing particular analytical tasks such as making comparisons or determining causality. The design principle of the information graphic should support the analytical task, showing the comparison or causality.

In his book The Visual Display of Quantitative Information, Edward Tufte defines graphical displays and principles for effective graphical display. He holds that excellence in statistical graphics consists of complex ideas communicated with clarity, precision and efficiency. Graphical displays should:

- show the data;
- induce the viewer to think about the substance rather than about methodology, graphic design, the technology of graphic production or something else;
- avoid distorting what the data has to say;
- present many numbers in a small space;
- make large data sets coherent;
- encourage the eye to compare different pieces of data;
- reveal the data at several levels of detail, from a broad overview to the fine structure;
- serve a reasonably clear purpose: description, exploration, tabulation or decoration; and
- be closely integrated with the statistical and verbal descriptions of a data set.

Indeed graphics can be more precise and revealing than conventional statistical computations.

Not applying these principles may result in misleading graphs, which distort the message or support an erroneous conclusion. Needlessly separating, the explanatory key from the image itself requires the eye to travel back and forth from the image to the key.

The Congressional Budget Office summarized several best practices for graphical displays in a June 2014 presentation. These included:

- Knowing your audience;
- Designing graphics that can stand alone outside the context of the report; and
- Designing graphics that communicate the key messages in the report.

3. Quantitative messages

Author Stephen Few describes eight types of quantitative messages that users may attempt to understand or communicate from a set of data and the associated graphs used to help communicate the message:

- Time-series: A single variable is captured over a period of time, such as the unemployment rate over a 10-year period. A line chart may be used to demonstrate the trend (see Figure 11-2).

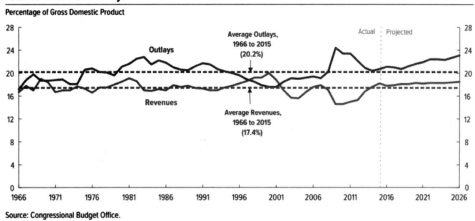

Figure 11-2 A time series illustrated with a line chart demonstrating trends in U.S. federal spending and revenue over time.

- Ranking: Categorical subdivisions are ranked in ascending or descending order, such as a ranking of sales performance (the measure) by sales persons (the category, with each sales person a categorical subdivision) during a single period. A bar chart may be used to show the comparison across the sales persons.
- Part-to-whole: Categorical subdivisions are measured as a ratio to the whole (i.e., a percentage out of 100%). A pie chart or bar chart can show the comparison of ratios, such as the market share represented by competitors in a market.
- Deviation: Categorical subdivisions are compared against a reference, such as a comparison of actual vs. budget expenses for several departments of a business for a given time period. A bar chart can show comparison of the actual versus the reference amount.
- Frequency distribution: Shows the number of observations of a particular variable for given intervals, such as the number of years in which the stock market return is between intervals such as 0-10%, 11%-20%, etc. A histogram, a type of bar chart, may be used for this analysis. A boxplot helps visualize key statistics about the distribution, such as median, quartiles, outliers, etc.
- Correlation: Comparison between observations represented by two variables (X,Y) to determine if they tend to move in the same or opposite directions. For example, plotting unemployment (X) and inflation (Y) for a sample of months. A scatter plot is typically used for this message (see Figure 11-3).

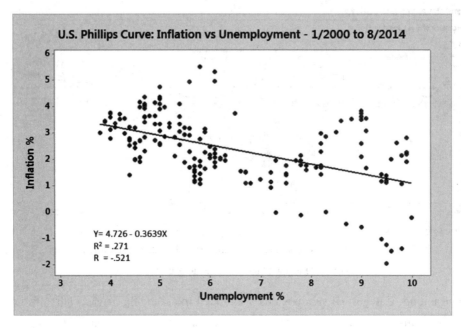

Figure 11-3　A scatter plot illustrating negative correlation between two variables (inflation and unemployment) measured at points in time.

- Nominal comparison: Comparing categorical subdivisions in no particular order, such as the sales volume by product code. A bar chart may be used for this comparison.
- Geographic or geospatial: Comparison of a variable across a map or layout, such as the unemployment rate by state or the number of persons on the various floors of a building. A cartogram is a typical graphic used.

Analysts reviewing a set of data may consider whether some or all of the messages and graphic types above are applicable to their task and audience. The process of trial and error to identify meaningful relationships and messages in the data is part of exploratory data analysis.

4. Visual Perception and Data Visualization

A human can distinguish differences in line length, shape, orientation, and color (hue) readily without significant processing effort; these are referred to as "pre-attentive attributes." For example, it may require significant time and effort (attentive processing) to identify the number of times the digit "5" appears in a series of numbers; but if that digit is different in size, orientation, or color, instances of the digit can be noted quickly through pre-attentive processing.

Effective graphics take advantage of pre-attentive processing and attributes and the relative strength of these attributes. For example, since humans can more easily process differences in line length than surface area, it may be more effective to use a bar chart (which takes advantage of line length to show comparison) rather than pie charts (which use surface area to show comparison).

Almost all data visualizations are created for human consumption. Knowledge of human perception and cognition is necessary when designing intuitive visualizations. Cognition refers to processes in human beings like perception, attention, learning, memory, thought, concept formation, reading, and problem solving. Human visual processing is efficient in detecting changes and making comparisons between quantities, sizes, shapes and variations in lightness. When properties of symbolic data are mapped to visual properties, humans can browse through large amounts of data efficiently. It is estimated that 2/3 of the brain's neurons can be involved in visual processing. Proper visualization provides a different approach to show potential connections, relationships, etc. which are not as obvious in non-visualized quantitative data. Visualization can become a means of data exploration.

5. Terminology

Data visualization involves specific terminology, some of which is derived from statistics. For example, author Stephen Few defines two types of data, which are used in combination to support a meaningful analysis or visualization:
- Categorical: Text labels describing the nature of the data, such as "Name" or "Age". This term also covers qualitative (nonnumerical) data.
- Quantitative: Numerical measures, such as "25" to represent the age in years.

Two primary types of information displays are tables and graphs.

A table contains quantitative data organized into rows and columns with categorical labels. It is primarily used to look up specific values. In the example above, the table might have categorical column labels representing the name (a qualitative variable) and age (a quantitative variable), with each row of data representing one person (the sampled experimental unit or category subdivision).

A graph is primarily used to show relationships among data and portrays values encoded as visual objects (e.g., lines, bars, or points). Numerical values are displayed within an area delineated by one or more axes. These axes provide scales (quantitative and categorical) used to label and assign values to the visual objects. Many graphs are also referred to as charts.

6. Data Presentation Architecture

Data presentation architecture (DPA) is a skill-set that seeks to identify, locate, manipulate, format and present data in such a way as to optimally communicate meaning and proper knowledge.

Historically, the term data presentation architecture is attributed to Kelly Lautt. Data Presentation Architecture (DPA) is a rarely applied skill set critical for the success and value of Business Intelligence. DPA is neither an IT nor a business skill set but exists as a separate field of expertise. Often confused with data visualization, data presentation architecture is a much broader skill set that includes determining what data on what schedule and in what exact format is to be presented, not just the best way to present data that has already been chosen. Data visualization skills are one element of DPA.

6.1 Objectives

DPA has two main objectives:
- To use data to provide knowledge in the most efficient manner possible (minimize noise, complexity, and unnecessary data or detail given each audience's needs and roles).
- To use data to provide knowledge in the most effective manner possible (provide relevant, timely and complete data to each audience member in a clear and understandable manner that conveys important meaning, is actionable and can affect understanding, behavior and decisions).

6.2 Scope

With the above objectives in mind, the actual work of data presentation architecture consists of:
- Creating effective delivery mechanisms for each audience member depending on their role, tasks, locations and access to technology
- Defining important meaning (relevant knowledge) that is needed by each audience member in each context
- Determining the required periodicity of data updates (the currency of the data)
- Determining the right timing for data presentation (when and how often the user needs to see the data)
- Finding the right data (subject area, historical reach, breadth, level of detail, etc.)
- Utilizing appropriate analysis, grouping, visualization, and other presentation formats

6.3 Related fields

DPA work shares commonalities with several other fields, including:
- Business analysis in determining business goals, collecting requirements, mapping processes.

- Business process improvement in that its goal is to improve and streamline actions and decisions in furtherance of business goals.
- Data visualization in that it uses well-established theories of visualization to add or highlight meaning or importance in data presentation.
- Graphic or user design: As the term DPA is used, it falls just short of design in that it does not consider such detail as colour palates, styling, branding and other aesthetic concerns, unless these design elements are specifically required or beneficial for communication of meaning, impact, severity or other information of business value. For example:

    (1) choosing locations for various data presentation elements on a presentation page (such as in a company portal, in a report or on a web page) in order to convey hierarchy, priority, importance or a rational progression for the user is part of the DPA skill-set;

    (2) choosing to provide a specific colour in graphical elements that represent data of specific meaning or concern is part of the DPA skill-set
- Information architecture, but information architecture's focus is on unstructured data and therefore excludes both analysis (in the statistical/data sense) and direct transformation of the actual content (data, for DPA) into new entities and combinations.

✎ New Words

discipline	['disiplin]	n.学科
		v.训练
variable	['vɛəriəbl]	n.变数，可变物，变量
		adj.可变的，不定的，易变的，变量的
plot	[plɔt]	n.图
		v.绘图
dot	[dɔt]	n.点，圆点
		vt.在……上打点
evidence	['evidəns]	n.明显，显著，明白，迹象，根据
usable	['ju:zəbl]	adj.可用的，便于使用的
causality	[kɔ:'zæliti]	n.因果关系
measurement	['meʒəmənt]	n.测量法，度量，（量得的）尺寸；度量单位制
chart	[tʃɑ:t]	n.图表
		vt.制图
sensor	['sensə]	n.传感器
aesthetic	[i:s'θetik]	adj.美学的，审美的

sparse	[spɑːs]	adj.稀少的，稀疏的
attention	[əˈtenʃən]	n.注意力，关注度
millennium	[miˈleniəm]	n.千年
precision	[priˈsiʒən]	n.精确，精密度，精度
distort	[disˈtɔːt]	vt.歪曲（事实等），误报
coherent	[kəuˈhiərənt]	adj.一致的，连贯的
reveal	[riˈviːl]	vt.展现，显示，揭示
tabulation	[ˌtæbjuˈleiʃən]	n.作表，表格
decoration	[ˌdekəˈreiʃən]	n.修饰，装饰；装饰品
verbal	[ˈvəːbəl]	adj.口头的
precise	[priˈsais]	adj.精确的，准确的 n.精确
misleading	[misˈliːdiŋ]	adj.易误解的，令人误解的
chartjunk	[ˈtʃɑːtdʒʌŋk]	n.垃圾图表
extraneous	[eksˈtreinjəs]	adj.无关系的
interior	[inˈtiəriə]	adj.内部的，内的 n.内部
gratuitous	[grəˈtjuːitəs]	adj.没必要的，无理由的
explanatory	[iksˈplænətəri]	adj.说明的，解释性的
debris	[ˈdebriː]	n.碎片，残骸
unemployment	[ˌʌnimˈplɔimənt]	n.失业，失业人数
deviation	[ˌdiːviˈeiʃən]	n.偏差，背离
interval	[ˈintəvəl]	n.间隔，距离 n.时间间隔
histogram	[ˈhistəugræm]	n.柱状图
quartile	[ˈkwɔːtail]	n.四分位数
outlier	[ˈautlaiə]	n.离群值，异常值
inflation	[inˈfleiʃən]	n.通货膨胀，物价暴涨
cartogram	[ˈkɑːtəgræm]	n.统计地图，变形地图；属性地图
perception	[pəˈsepʃən]	n.理解，感知，感觉
hue	[hjuː]	n.色调，颜色，色彩
cognition	[kɔgˈniʃən]	n.认识
neuron	[ˈnjuərɔn]	n.神经细胞，神经元
combination	[ˌkɔmbiˈneiʃən]	n.结合，联合，合并
categorical	[ˌkætiˈgɔrikəl]	adj.分类的，按类别的；无条件的，绝对的
qualitative	[ˈkwɔlitətiv]	adj.定性的，性质上的
nonnumerical	[ˈnɔnnjuˈmerikəl]	adj.非数值的

quantitative	['kwɔntitətiv]	adj.数量的；定量的
subdivision	['sʌbdi,viʒən]	n.细分，一部
portray	[pɔː'trei]	v.描绘，描述
delineate	[di'linieit]	v.绘画，勾画
axes	['æksiːz]	n.轴
scale	[skeil]	n.刻度，衡量，比例，数值范围，等级
intelligence	[in'telidʒəns]	n.智力，智能
expertise	[,ekspə'tiːz]	n.专家的意见，专门技术
confused	[kən'fjuːzd]	adj.困惑的，烦恼的
exact	[ig'zækt]	adj.精确的，准确的
complexity	[kəm'pleksiti]	n.复杂性；复杂的事物
unnecessary	[ʌn'nesisəri]	adj.不必要的，多余的
role	[rəul]	n.角色，任务
convey	[kən'vei]	vt.传达，转让
periodicity	[,piəriə'disiti]	n.周期
grouping	['gruːpiŋ]	n.分组
improvement	[im'pruːvmənt]	n.改进，进步
streamline	['striːmlain]	v.使现代化；精简，简化；使成流线型
severity	[si'veriti]	n.严肃，严格，严重，激烈
portal	['pɔːtəl]	n.入口，门户
hierarchy	['haiərɑːki]	n.层次
priority	[prai'ɔriti]	n.优先，优先权
rational	['ræʃənl]	adj.理性的，合理的，推理的
exclude	[iks'kluːd]	vt.拒绝接纳，排斥

🐾 Phrases

data visualization	数据可视化
visual communication	视觉传达，视觉传播
visual representation	直观表示
statistical graphics	统计图，统计图形学
design principle	设计原理
descriptive statistics	描述统计学
grounded theory	扎根理论
Internet of things	物联网
data analysis	数据分析
data science	数据科学

hand in hand	携手，手拉手；密切合作
data set	数据集
fail to	未能……
broad overview	宏观视角，概览
fine structure	精细结构
integrate with …	使与……结合
Congressional Budget Office	美国会预算办公室
line chart	线形图，线图，线形图表
categorical subdivision	类别细分
ascending order	升序
descending order	降序
bar chart	柱状图
pie chart	饼图
market share	市场份额
frequency distribution	频率分布
opposite direction	反向，相反方向
scatter plot	散点图
trial and error	反复试验
pre-attentive attribute	前注意属性
be involved in	涉及，专心
be derived from	源自于
text label	文本标签，文字标签；文字标记，文本标号
numerical measure	数字型度量
visual object	视觉对象，视频对象
Periodic Table of Visualization Methods	可视化方法周期表
business intelligence	商业智能
business analysis	商业分析
colour palate	调色板
web page	网页

Abbreviations

DPA (Data Presentation Architecture)	数据呈现结构
IT (Information Technology)	信息技术

✎ Notes

[1] Tables are generally used when users will look up a specific measurement, while charts of various types are used to show patterns or relationships in the data for one or more variables.

本句中，while 是一个连词，连接两个并列的句子，表示对比，意思是"而"。when users will look up a specific measurement 是一个时间状语从句，修饰谓语 are generally used。to show patterns or relationships in the data for one or more variables 是一个动词不定式短语，作目的状语。

[2] To convey ideas effectively, both aesthetic form and functionality need to go hand in hand, providing insights into a rather sparse and complex data set by communicating its key-aspects in a more intuitive way.

本句中，To convey ideas effectively 是一个动词不定式短语，作目的状语。providing insights into a rather sparse and complex data set by communicating its key-aspects in a more intuitive way 是一个现在分词短语，作结果状语。

[3] Not applying these principles may result in misleading graphs, which distort the message or support an erroneous conclusion.

本句中，Not applying these principles 是一个动名词短语，作主语。which distort the message or support an erroneous conclusion 是一个非限定性定语从句，对宾语 misleading graphs 进行补充说明。result in 的意思是"导致"。

[4] For example, since humans can more easily process differences in line length than surface area, it may be more effective to use a bar chart (which takes advantage of line length to show comparison) rather than pie charts (which use surface area to show comparison).

本句中，since humans can more easily process differences in line length than surface area 是一个原因状语从句。(which takes advantage of line length to show comparison)是一个定语从句，修饰和限定 a bar chart。(which use surface area to show comparison)也是一个定语从句，修饰和限定 pie charts。

[5] Often confused with data visualization, data presentation architecture is a much broader skill set that includes determining what data on what schedule and in what exact format is to be presented, not just the best way to present data that has already been chosen.

本句中，that includes determining what data on what schedule and in what exact format is to be presented 是一个定语从句，修饰和限定 a much broader skill set。that has already been chosen 也是一个定语从句，修饰和限定 data。

Exercises

【Ex. 1】 根据课文内容回答问题。

1. What is data visualization viewed by many disciplines as?
2. What is a primary goal of data visualization?
3. What does data visualization refer to?
4. What are the several best practices for graphical displays the Congressional Budget Office summarized in a June 2014 presentation?
5. What are the eight types of quantitative messages Author Stephen Few describes that users may attempt to understand or communicate from a set of data and the associated graphs used to help communicate the message?
6. What does cognition refer to?
7. What is efficient in detecting changes and making comparisons between quantities, sizes, shapes and variations in lightness?
8. What are the two primary types of information displays mentioned in the passage? What are they primarily used to respectively?
9. What is data presentation architecture (DPA)?
10. How many main objectives does DPA have? What are they?

【Ex. 2】 把下列句子翻译为中文。

1. Each input parameter should have the variable name and its value.
2. If a computer user fails to log off, the system is accessible to all.
3. User experience designers are great at making software friendly and usable for new customers.
4. There are no previous statistics for comparison.
5. On modern hardware and operating systems, it can deliver accuracy and precision in the microsecond range.
6. A histogram is used to graphically summarize and display the distribution of a process data set.
7. You will need hardware, software, and network expertise.
8. This considerably reduces the debugging time and complexity.
9. In addition, unnecessary processing time and resources are being consumed starting and stopping the transaction.
10. This may have been an improvement, but "breakthrough" was an overstatement.

【Ex. 3】 短文翻译。

7 Important Types of Big Data

Big data is a term thrown around in a lot of articles, and for those who understand what big data means that is fine, but for those struggling to understand exactly what big data is, it can get frustrating. There are several definitions of big data as it is frequently used as an all-encompassing term for everything from actual data sets to big data technology and big data analytics. However, this article will focus on the actual types of data that are contributing to the ever growing collection of data referred to as big data. Specifically we focus on the data created outside of an organization, which can be grouped into two broad categories: structured and unstructured.

1. Structured Data

1.1 Created

Created data is just that; data businesses purposely create, generally for market research. This may consist of customer surveys or focus groups. It also includes more modern methods of research, such as creating a loyalty program that collects consumer information or asking users to create an account and login while they are shopping online.

1.2 Provoked

A Forbes Article defined provoked data as, "Giving people the opportunity to express their views." Every time a customer rates a restaurant, an employee, a purchasing experience or a product they are creating provoked data. Rating sites, such as Yelp, also generate this type of data.

1.3 Transacted

Transactional data is also fairly self-explanatory. Businesses collect data on every transaction completed, whether the purchase is completed through an online shopping cart or in-store at the cash register. Businesses also collect data on the steps that lead to a purchase online. For example, a customer may click on a banner ad that leads them to the product pages which then spurs a purchase.

As explained by the Forbes article, "Transacted data is a powerful way to understand exactly what was bought, where it was bought, and when. Matching this type of data with other information, such as weather, can yield even more insights.

1.4 Compiled

Compiled data is giant databases of data collected on every U.S. household. Companies like Acxiom collect information on things like credit scores, location, demographics, purchases and registered cars that marketing companies can then access for supplemental consumer data.

1.5 Experimental

Experimental data is created when businesses experiment with different marketing pieces and messages to see which are most effective with consumers. You can also look at experimental data as a combination of created and transactional data.

2. Unstructured Data

People in the business world are generally very familiar with the types of structured data mentioned above. However, unstructured is a little less familiar not because there's less of it, but before technologies like NoSQL and Hadoop came along, harnessing unstructured data wasn't possible. In fact, most data being created today is unstructured. Unstructured data, as the name suggests, lacks structure. It can't be gathered based on clicks, purchases or a barcode, so what is it exactly?

2.1 Captured

Captured data is created passively due to a person's behavior. Every time someone enters a search term on Google that is data that can be captured for future benefit. The GPS info on our smartphones is another example of passive data that can be captured with big data technologies.

2.2 User-generated

User-generated data consists of all of the data individuals are putting on the Internet every day. From tweets, to Facebook posts, to comments on news stories, to videos put up on YouTube, individuals are creating a huge amount of data that businesses can use to better target consumers and get feedback on products.

Big data is made up of many different types of data. The seven listed above comprise types of external data included in the big data spectrum. There are, of course, many types of internal data that contribute to big data as well, but hopefully breaking down the types of data helps you to better see why combining all of this data into big data is so powerful for business.

【Ex. 4】 将下列词填入适当的位置（每词只用一次）。

| track | sophisticated | correlations | databases | insights |
| visualization | algorithms | accumulated | manipulate | environments |

Data Visualization

Data visualization is a general term that describes any effort to help people understand the significance of data by placing it in a visual context. Patterns, trends and correlations that might go undetected in text-based data can be exposed and recognized easier with data visualization software.

Today's data visualization tools go beyond the standard charts and graphs used in Microsoft Excel spreadsheets, displaying data in more __(1)__ ways such as infographics, dials and gauges, geographic maps, sparklines, heat maps, and detailed bar, pie and fever charts. The images may include interactive capabilities, enabling users to __(2)__ them or drill into the data for querying and analysis. Indicators designed to alert users when data has been updated or predefined conditions occur can also be included.

1. Importance of data visualization

Data visualization has become the de facto standard for modern business intelligence(BI). The success of the two leading vendors in the BI space, Tableau and Qlik —both of which heavily emphasize __(3)__ — has moved other vendors toward a more visual approach in their software. Virtually all BI software has strong data visualization functionality.

Data visualization tools have been important in democratizing data and analytics and making data-driven __(4)__ available to workers throughout an organization. They are typically easier to operate than traditional statistical analysis software or earlier versions of BI software. This has led to a rise in lines of business implementing data visualization tools on their own, without support from IT.

Data visualization software also plays an important role in big data and advanced analytics projects. As businesses __(5)__ massive troves of data during the early years of the big data trend, they needed a way to quickly and easily get an overview of their data. Visualization tools were a natural fit.

Visualization is central to advanced analytics for similar reasons. When a data scientist is writing advanced predictive analytics or machine learning __(6)__, it becomes important to visualize the outputs to monitor results and ensure that models are performing as intended. This is because visualizations of complex algorithms are generally easier to interpret than numerical outputs.

2. Examples of data visualization

Data visualization tools can be used in a variety of ways. The most common use today is as a BI reporting tool. Users can set up visualization tools to generate automatic dashboards that track company performance across key performance indicators and visually interpret the results.

Many business departments implement data visualization software to __(7)__ their own initiatives. For example, a marketing team might implement the software to monitor the performance of an email campaign, tracking metrics like open rate, click-through rateand conversion rate.

As data visualization vendors extend the functionality of these tools, they are increasingly being used as front ends for more sophisticated big data __(8)__. In this setting, data visualization software helps data engineers and scientists keep track of data sources and do basic exploratory analysis of data sets prior to or after more detailed advanced analyses.

3. How data visualization works

Most of today's data visualization tools come with connectors to popular data sources, including the most common relational __(9)__, Hadoop and a variety of cloud storage platforms. The visualization software pulls in data from these sources and applies a graphic type to the data.

Data visualization software allows the user to select the best way of presenting the data, but, increasingly, software automates this step. Some tools automatically interpret the shape of the data and detect __(10)__ between certain variables and then place these discoveries into the chart type that the software determines is optimal.

Typically, data visualization software has a dashboard component that allows users to pull multiple visualizations of analyses into a single interface, generally a web portal.

Text B

The 14 Best Data Visualization Tools

Raw data is boring and it's difficult to make sense of it in its natural form. Add visualization to it and you get something that everybody can easily digest. You can not only make sense of it faster, but also observe interesting patterns that wouldn't be apparent from looking only at stats.

To make the tedious task of making beautiful charts and maps easier, I've made the list of best data visualization tools available for the job. I've divided the list into two parts; first covers the tools that require coding and are meant for developers, while the second list contains data visualization software products that don't require any coding.

Let's get started!

1. For Developers

1.1 D3.js

D3.js, short for "Data Driven Documents", is the first name that comes to mind when we think of a Data Visualization Software. It uses HTML, CSS, and SVG to render some amazing charts and diagrams. If you can imagine any visualization, you can do it with D3. It is feature packed, interactivity rich and extremely beautiful. Most of all, it's free and open-source.

It doesn't ship with pre-built charts out of the box, but has a nice gallery which showcases what's possible with D3. There are two major concerns with D3.js: it has a steep learning curve and it is compatible only with modern browsers (IE 9+). So pick it up only when you have enough time in hand and are not concerned about displaying your charts on older browsers.

1.2 FusionCharts

FusionCharts has probably the most exhaustive collection of charts and maps. With over 90+ chart types and 965 maps, you'll find everything that you need right out of the box. It not only supports modern browsers, but also older browsers starting from IE 6.

FusionCharts supports both JSON and XML data formats, and you can export charts in PNG, JPEG, SVG or PDF. They have a nice collection of business dashboards and live demos for inspiration.

Their charts and maps work across all devices and platforms, are highly customizable and have beautiful interactions. One thing to keep in mind about FusionCharts is that it's slightly expensive. But you can always get started with their unrestricted free trial and then buy if you like it.

1.3 Chart.js

Chart.js is a tiny open source library that supports just six chart types: line, bar, radar, polar, pie and doughnut. But the reason I like it is that sometimes that's all the charts one needs for a project. If the application is big and complex, then libraries like Google Charts and FusionCharts makes sense, otherwise for small hobby projects Chart.js is the perfect solution.

It uses HTML5 canvas element for rendering charts. All the charts are responsive and use flat design. It is one of the most popular open-source charting libraries to emerge recently. Check out the documentation for live examples of all six chart types.

1.4　Google Charts

Google Charts renders charts in HTML5/SVG to provide cross-browser compatibility and cross-platform portability to iPhones and Android. It also includes VML for supporting older IE versions.

It offers a decent number of charts which covers the most commonly used chart types like bar, area, pie and gauges. It is flexible and user friendly (because Google!). You can view this gallery to get an idea of various charts and the type of interactions to expect.

1.5　Highcharts

Highcharts is another big player in the charting space. Like FusionCharts, it also offers a diverse range of charts and maps right out of the box. Other than normal charts, it also offers a different package for stock charts called Highstock which is also feature rich.

It allows exporting charts in PNG, JPG, SVG and PDF. You can view the various chart types it offers in the demo section. Highcharts is free for non-commercial and personal use, but you will have to buy a license for deploying it in commercial applications.

1.6　Leaflet

Leaflet is an open-source library developed by Vladimir Agafonkin for mobile-friendly interactive maps. It is extremely light (at just 33kb) and has lots of features for making any kind of maps. It uses HTML5 and CSS3 for rendering maps, and works across all major desktop and mobile platforms. In the words of Vladimir Agafonkin, Leaflet is designed with simplicity, performance and usability in mind.

There is a wide range of plugins available for adding features like animated markers, heatmaps etc. that extend the core functionality. If you are thinking of developing an application that involves maps, you should give Leaflet a try.

1.7　Dygraphs

Dygraphs is an open-source JavaScript charting library for handling huge data sets. It's fast, flexible and highly customizable. It works in all major browsers (including IE8) and has an active community.

Dygraphs has defined a niche use case for itself and won't be the perfect solution for all your needs. But it will work for you more often than not whenever you are handling large datasets. To explore what is possible, check out this nicely designed demo gallery.

2. Non-Developers

2.1 Datawrapper

Datawrapper is an online tool for making interactive charts. Once you upload the data from CSV file or paste it directly into the field, Datawrapper will generate a bar, line or any other related visualization. Many reporters and news organizations use Datawrapper to embed live charts into their articles. It is very easy to use and produces effective graphics.

2.2 Tableau

Tableau Public is perhaps the most popular visualization tool which supports a wide variety of charts, graphs, maps and other graphics. It is a completely free tool and the charts you make with it can be easily embedded in any web page. They have a nice gallery which displays visualizations created via Tableau.

Although it offers charts and graphics that are much better than other similar tools, I don't 'love' to use its free version because of the big footer it comes with. If it's not as big a turn-off for you as it is for me, then you should definitely give it a try. Or if you can afford it, you can go for a paid version.

2.3 Raw

Raw defines itself as the missing link between spreadsheets and vector graphics. It is built on top of D3.js and is extremely well designed. It has such an intuitive interface that you'll feel like you've used it before. It is open-source and doesn't require any registration.

It has a library of 16 chart types to choose from and all the processing is done in browser. So your data is safe. RAW is highly customizable and extensible, and can even accept new custom layouts.

2.4 Timeline JS

As the name suggests, Timeline JS helps you create beautiful timelines without writing any code. It is a free, open-source tool which is used by some of the most popular websites like Time and Radiolab.

It's very easy to follow four-step process to create your timeline which is explained here. Best part? It can pull in media from a variety of sources and has built-in support for Twitter,

Flickr, Google Maps, YouTube, Vimeo, Vine, Dailymotion, Wikipedia, SoundCloud and other similar sites.

2.5 Infogram

Infogram enables you to create both charts and infographics online. It has a restricted free version and two paid options which include features like 200+ maps, private sharing and icons library etc.

It comes with an easy-to-use interface and its basic charts are well designed. One feature that I don't like is the huge logo that you get when you try to embed interactive charts into your webpage (in free version). It will be better if they can make it like the little text that Datawrapper uses.

2.6 Plotly

Plotly is a web-based data analysis and graphing tool. It supports a good collection of chart types with built in social sharing features. The charts and graph types available have a professional look and feel. Creating a chart is just a matter of loading in your information and customizing the layout, axes, notes and legend. If you are looking to get started, you can find some inspiration here.

2.7 ChartBlocks

ChartBlocks is another online chart builder that is well designed and allows you to build basic charts very quickly. It has a limited number of chart types, but that will not be a problem as most common chart types are covered.

It allows you to pull in data from multiple external sources like spreadsheets and databases. After you have made the chart, you can either export it via SVG or PNG, embed it in your website or share it on social media.

✎ New Words

digest	[dai'dʒest]	vt.消化，理解；融会贯通；分类；整理
	['daidʒest]	n.分类，摘要
stats	[stæts]	n.统计学，统计表（=statistics）
tedious	['tiːdiəs]	adj.沉闷的，冗长乏味的
diagram	['daiəgræm]	n.图表
gallery	['gæləri]	n.图库
showcase	['ʃəukeis]	n.（商店或博物馆的玻璃）陈列橱
exhaustive	[ig'zɔːstiv]	adj.无遗漏的，彻底的，详尽的

dashboard	[ˈdæʃˌbɔːd]	n.仪表板
demo	[ˈdeməu]	n.演示
inspiration	[ˌinspəˈreiʃən]	n.灵感
customizable	[ˈkʌstəmaizəbl]	n.用户化，专用化，定制
doughnut	[ˈdəunʌt]	n.圆环图
hobby	[ˈhɔbi]	n.业余爱好
responsive	[risˈpɔnsiv]	adj.响应的
compatibility	[kəmˌpætiˈbiliti]	n.兼容性
gauge	[gedʒ]	n.标准尺，量规，量表
		n.测量
usability	[ˌjuːzəˈbiləti]	n.可用性
heatmap	[hiːtmæp]	n.热图
intuitive	[inˈtjuːitiv]	adj.直觉的
registration	[ˌredʒisˈtreiʃən]	n.注册，登记
extensible	[ikˈstensibl]	adj.可扩展的，可延长的
timeline	[ˈtaimlain]	n.时间轴，时间线
restricted	[risˈtriktid]	adj.受限制的，有限的

✎ Phrases

make sense of…	搞清……的意思
Data Driven Documents	数据驱动的文档
steep learning curve	陡峭的学习曲线
be compatible with	适合，一致
pick up on …	与……熟悉起来
be concerned about	关心，挂念
free trial	免费试用
canvas element	画布元素
rendering chart	渲染图
flat design	扁平化设计
stock chart	股票图
animated marker	动画制作器，动画制作程序
use case	用例
more often than not	往往，多半
web page	网页
turn-off	使人扫兴（或倒胃口）的事物
a variety of	多种的

social sharing　　　　　　　　　社交分享
limited number　　　　　　　　　少数
social media　　　　　　　　　　社交媒体

Abbreviations

CSS (Cascading Style Sheets)　　　　　　层叠样式表
SVG (Scalable Vector Graphics)　　　　　指可伸缩矢量图形
JSON (JavaScript Object Notation)　　　　JS 对象标记
XML (eXtensible Markup Language)　　　可扩展标记语言
PNG (Portable Network Graphic)　　　　 便携式网络图像
JPEG (Joint Photographic Experts Group)　联合图像专家小组
PDF (Portable Document Format)　　　　 便携式文档格式
CSV (Comma-Separated Values)　　　　　逗号分隔值，字符分隔值

Exercises

【Ex. 5】 根据课文内容填空。

1. D3.js is short for _____. It is the first name that comes to mind when we think of a Data Visualization Software. It uses _____, _____, and _____ to render some amazing charts and diagrams.

2. FusionCharts supports both _____ and _____ data formats, and you can export charts in PNG, JPEG, SVG or PDF.

3. Chart.js is a tiny _____ that supports just six chart types: _____, _____, radar, polar, _____ and doughnut.

4. Google Charts offers a decent number of charts which covers the most commonly used chart types like _____, _____, _____ and _____.

5. Dygraphs is an open-source _____ for handling huge data sets. It's fast, _____ and highly _____. It works in all major browsers (including IE8) and has _____.

6. Datawrapper is an online tool for making _____. Once you upload the data from CSV file or paste it directly into _____, Datawrapper will generate a bar, line or any other _____.

7. Tableau Public is perhaps the most popular visualization tool which supports a wide variety of _____, _____, _____ and other graphics.

8. Raw defines itself as the missing link between _____ and _____. It is built on top of D3.js and is extremely well designed. It has a library of _____ to choose

from and all the processing is done in _____.
9. Infogram enables you to create both charts and infographics _____. It has a _____ and two paid options which include features like _____, _____ and icons library etc.
10. Plotly is a _____ data analysis and graphing tool. It supports a good collection of chart types with _____.

参考译文

数据可视化

数据可视化被许多学科视为视觉传达的现代方式。它涉及创建和研究数据的视觉表示，意味着"信息已经抽象成一些图像形式，包括信息单元的属性或变量"。

数据可视化的主要目标是通过统计图形、绘图和信息图形清晰有效地传达信息。可以使用点、线或条来对数字数据进行编码，以可视化地传达定量消息。有效的可视化帮助用户分析和理解数据和证据。它使复杂的数据更易于访问、可理解和可用。用户可以执行特定的分析任务，例如进行比较或理解因果关系以及图形的设计原理（即，显示比较或显示因果关系）。用户通常使用表格来查找特定的度量，而各种类型的图表用于显示数据中的模式以及一个或多个变量的关系。

数据可视化既是艺术又是科学。一些人将其视为描述性统计的一个分支，其他人视其为扎根理论开发工具。互联网活动和越来越多的环境传感器制造的数据越来越多。处理、分析和传达这些数据是数据可视化所面临的伦理和分析挑战。数据科学家帮助解决了这一挑战。

1. 概述

数据可视化是指用于传达数据或信息的技术，它通过将数据或信息编码为图形中的视觉对象（例如点、线或条）来实现。其目标是向用户清楚有效地传达信息。它是数据分析或数据科学的步骤之一。根据 Friedman 的说法，数据可视化的主要目标是通过图形手段清晰有效地传达信息，并不意味着数据可视化需要看起来很无聊但有用或者看起来很漂亮但很复杂。为了有效地传递观点，需要兼顾美学形式和功能，能洞察极为稀少和复杂的数据集，以更直观的方式传达其关键方面。但设计者往往无法在形式和功能之间取得平衡，创造华丽的数据可视化形式并不能满足其主要目的——传达信息。

事实上，Fernanda Viegas 和 Martin M. Wattenberg 建议，理想的可视化应该不仅要清楚地沟通，而且可以激发观众的参与和关注。

数据可视化与信息图形、信息可视化、科学可视化、探索性数据分析和统计图形密切相关。在新千年中，数据可视化已成为研究、教学和发展的活跃领域。

图 11-1 数据可视化是分析数据并将其呈现给用户的步骤之一（图略）。

2. 有效图形显示的特点

爱德华·图夫特（Edward Tufte）教授解释说，信息显示的用户正在执行特定的分析任务，如进行比较或确定因果关系。信息图形的设计原则应该支持分析任务、显示比较或因果关系。

在《定量信息视觉显示》一书中，爱德华·图夫特定义了"图形显示"和有效显示图形原理，他认为优秀的统计图形包括清晰、精确和有效地传达复杂思想。图形显示应该：

- 显示数据；
- 引导观众思考本质，而不是只关注方法论、图形设计、图形制作技术等；
- 避免扭曲数据所说的内容；
- 在小的空间内显示多的数字；
- 使大数据集一致；
- 鼓励用眼睛去比较不同的数据块；
- 从多个层面细述数据，从概述到细微结构；
- 提供相当明确的目的：描述、探索、制表或装饰；
- 把数据集的统计和口头描述紧密结合。

与传统的统计计算相比，图形可以更精确并更有启发性。

不采用这些原则可能会导致误导性图表，从而扭曲信息或支持错误的结论。不要把说明性的关键词与图像本身分开，那样会要求眼球在图像与关键点之间来回移动。

国会预算办公室在 2014 年 6 月的演示文稿中总结了图形显示的几种最佳做法。包括：

- 了解你的观众；
- 设计可以在报告背景之外独立的图形；
- 设计在报告中传达关键信息的图形。

3. 定量信息

作者 Stephen Few 描述了八种类型的定量消息，用户可以从用来传达信息的数据集和相关图形中试图理解这些信息，并能与之通信。

- 时间序列：捕获在一段时间内一个变量的值，如十年的失业率。线形图可用于展示其趋势。

- 排名：按升序或降序排列细分的类别，如单一时期销售人员（该类别，每个销售人员为细分类别）的销售业绩（度量）排名。可以使用条形图显示销售人员的比较。
- 部分到全部：以整体的比例（即100%的百分比）来细分类别。饼图或条形图可以显示比例的比较，例如，市场中竞争对手所代表的市场份额。
- 偏差：将细分类别与参考项进行比较，例如对于给定时间段的几个业务部门的实际费用与预算额加以比较。条形图可以显示实际量与参考量的比较。
- 频率分布：显示给定间隔的特定变量的观察次数，例如股票市场在0~10%、11%~20%等间隔之间的年数。一种称为直方图的条形图可用于此分析。boxplot可以帮助显示有关分布的关键统计信息，如中位数、四分位数及异常值等。
- 相关性：比较两个变量（X，Y）表示的观察值，以确定它们是否趋向于相同或相反的方向。例如，在几个月的样本中绘制失业（X）和通货膨胀（Y）的关系图。散布图通常用于表示此类消息。
- 名义比较：对没有特定顺序的细分类别加以比较，如按产品代码比较销售量。可以使用条形图进行此类比较。
- 地理或地理空间：地图或布局之间的变量比较，例如国家的失业率或建筑物各层楼的人数。直方图是常用的典型图形。

审查一组数据的分析师可能会考虑上述部分或全部消息和图形类型是否适用于其任务和受众。探索性数据分析就是在数据中识别有意义的关系和消息。

图11-2是一个时间序列说明线图，展示了美国联邦消费和收入随时间的趋势（图略）。

图11-3是一个散点图，显示了在时间点测量的两个变量（通货膨胀和失业）之间的负相关（图略）。

4. 视觉感知和数据可视化

人可以容易地区分线长、形状、方向和颜色（色调）上的差异，而不需要大量的处理工作；这些被称为"前注意属性"。例如，识别数字"5"出现在一系列数字中的次数可能需要大量的时间和精力（"注意处理"），但如果该数字的大小、方向或颜色不同，则可以通过前注意处理快速注意到该数字。

有效的图形利用了前注意处理和属性以及这些属性的相对强度。例如，由于人可以更容易地处理线路长度与表面积的差异，使用条形图（利用线长度来显示比较）可能比饼图（使用表面积来显示比较）更有效。

几乎所有的数据可视化都是为人类消费而创建的。在设计直觉可视化时，需要了解人的感知和认知。认知是指人的处理过程，如感知、注意力、学习、记忆、思维、概念形成、阅读和解决问题。人类视觉处理在检测变化方面是有效的，并且能在数量、大小、形状和亮度变化之间进行比较。当符号数据的属性映射为可视化属性时，人们可以有效地浏览大量的数据。据估计，大脑神经元的 2/3 可以参与视觉处理。正确的可视化提供了一种不同的方法来显示在非可视化定量数据中并不明显的潜在连接及关系等。可视化可以成为数据探索的一种手段。

5. 术语

数据可视化涉及特定术语，其中一些来自统计学。例如，作者 Stephen Few 定义了两种类型的数据，它们组合使用以支持有意义的分析或可视化。

- 分类：描述数据性质的文本标签，如"名称"或"年龄"。该术语还包括定性（非数值）数据。
- 定量：数字度量，如"25"代表年龄。

信息显示的两种主要类型是表格和图表。

- 表格包含按照分类标签组织成行和列的定量数据。它主要用于查找特定值。在上面的示例中，表格可能具有表示名称（定性变量）和年龄（定量变量）的分类列标签，每行数据表示一个人（抽样的实验单位或细分类别）。
- 图形主要用于显示编码为视觉对象（例如，线条或点）的数据和描绘值之间的关系。数值显示在由一个或多个轴描绘的区域内。这些轴提供了用于标记和分配视觉对象值的比例（定量和分类）。许多图也被称为图表。

6. 数据呈现结构

数据呈现结构（DPA）是一个技能集，旨在以适当的知识来识别、定位、操纵、格式化和呈现数据，并以最佳方式传达意义。

历史上，术语数据呈现结构归功于凯利·劳特（Kelly Lautt）。数据呈现结构（DPA）是一种很少应用的技能，对商业智能的成功和价值至关重要。DPA 既不是 IT，也不是业务技能，而是作为一个独立的专业领域存在，通常与数据可视化混淆，数据呈现结构是一个更广泛的技能，包括确定什么样的数据按照什么时间表以及何时提供准确的格式，不仅仅是用最佳方式呈现已经选择的数据。数据可视化技能是 DPA 的一个要素。

6.1 目标

DPA 有两个主要目标：

- 以最有效的方式使用数据提供知识（尽可能减少噪音、复杂性和不必要的数据或详细信息，以满足每个受众的需求和角色）。
- 以最有效的方式使用数据提供知识（以清晰易懂的方式为每个受众提供相关、及时和完整的数据，这个方式要有重要意义，可操作，可理解，并能够影响其行为和决策）。

6.2 范围

基于上述目标，数据呈现结构的实际工作包括：
- 根据其角色、任务、位置和访问技术，为每个受众成员创建有效的交付机制。
- 定义每个观众在每个环境中需要的重要意义（相关知识）。
- 确定所需的数据更新周期（数据的流通）。
- 确定数据呈现的正确时机（用户需要查看数据的时间和频率）。
- 查找正确的数据（主题区域、历史范围、广度、细节级别等）。
- 利用适当的分析、分组、可视化和其他呈现格式。

6.3 相关领域

DPA 也可用于其他几个领域，包括：
- 业务分析以确定业务目标、收集需求、过程图示。
- 业务流程改进，其目标是改进和简化行动和决策，促进实现业务目标。
- 数据可视化，它使用完善的可视化理论把数据的意义或重要性突出呈现出来。
- 图形或用户设计：使用 DPA 术语时，除非如色差、造型、品牌和其他美学细节特别需要、有益沟通或影响其商业价值，就不考虑这些设计元素。例如：

  （1）在演示页面（如公司门户、报告或网页）中选择各种数据表示元素的位置，以便为用户传达层次结构、优先级、重要性或合理的进展。这是 DPA 技能集的一部分。

  （2）选择在图形元素中提供特定颜色，表示特定意义或关注的数据，这也是 DPA 技能集的一部分。

- 信息架构，但信息架构的重点是非结构化数据，因此排除了（统计/数据意义上）的分析，并将实际内容（数据、DPA）直接转换为新的实体和组合。

Unit 12

Text A

How to Manage Big Data's Big Security Challenges

As the amount of data being collected continues to grow, more and more companies are building big data repositories to store, aggregate and extract meaning from their data. Big data provides an enormous competitive advantage for corporations, helping businesses tailor their products to consumer needs, identify and minimize corporate inefficiencies, and share data with user groups across the enterprise. With a growth rate of 58 percent in 2017 alone, these technologies and their benefits are here to stay.

Unfortunately, legitimate organizations aren't the only groups that are going big. Large sets of consolidated data are a tempting target for cyber attackers. Breaching an organization's big data repository can provide criminal groups with bigger payoffs. And when attackers set their sights on big data repositories, the effects can be devastating for the affected organizations. Terabytes of data in these repositories may include a company's crown jewels customer data, employee data, and trade secrets. The recent data breach at Target is estimated to cost the company upwards of $1.1 billion, and the PlayStation breach cost Sony an estimated $171 million. A breach in a big data repository could be even more damaging at a financial institution or healthcare provider, where the value of the data is extremely high and government regulations come into play.

1. The Data

The variety, velocity and volume of big data amplify the security management challenges that are addressed in traditional security management. Big data repositories will include information deposited by various sources across the enterprise. This variety of data makes secure access management a challenge. Each data source will have its own access restrictions and security policies, making it difficult to balance appropriate security for all data sources with the need to aggregate and extract meaning from the data. For example, a big data environment may include a dataset with proprietary research information, a dataset requiring regulatory compliance, and a separate dataset with personally identifiable information (PII). A researcher might want to correlate their research with a dataset including PII data, but what restrictions should be in-place to ensure adequate security? Protecting big data requires balancing analysis like this with security requirements on a case-by-case basis.

In addition, many of the repositories collect data at high volumes and velocity from a number of different data sources, and they all might have their own data transfer workflows. These connections to multiple repositories can increase the attack surface for an adversary. A big data system receiving feeds from 20 different data sources may present an attacker with 20 viable vectors to attempt to gain access to a cluster.

2. The Infrastructure

Another big data challenge is the distributed nature of big data environments. Compared with a single high-end database server, distributed environments are more complicated and vulnerable to attack. When big data environments are distributed geographically, physical security controls need to be standardized across all accessible locations. When data scientists across the organization want access to information, perimeter protection becomes important and complicated to ensure access to users while protecting the system from a possible attack. With a large number of servers, there is an increased possibility that the configuration of servers may not be consistent — and that certain systems may remain vulnerable.

3. The Technology

An additional big data security challenge is that big data programming tools, including Hadoop and NoSQL databases, were not originally designed with security in mind. For example, Hadoop originally didn't authenticate services or users, and didn't encrypt data that's transmitted between nodes in the environment. This creates vulnerabilities for authentication and network security. NoSQL databases lack some of the security features provided by traditional databases, such as role-based access control. The advantage of NoSQL is that it allows for the flexibility to include new data types on the fly, but defining security policies for this new data is not straightforward with these technologies.

4. Securing Big Data

So what can be done to help bring the security of traditional database management to big data? Several organizations describe and define different security controls. The SANS Institute provides a list of 20 security controls. The list contains several controls that I would recommend to address the security challenges presented by big data.

- Application Software Security. Use secure versions of open-source software. As described above, big data technologies weren't originally designed with security in mind. Using open-source technologies like Apache Accumulo or the .20.20x version of Hadoop or above can help address this challenge. In addition, proprietary technologies like Cloudera Sentry or DataStax Enterprise offer enhanced security at the application layer. Specifically, Sentry and Accumulo also support role-based access control to enhance security for NoSQL databases.

- Maintenance, Monitoring, and Analysis of Audit Logs. Implement audit logging technologies to understand and monitor big data clusters. Technologies like Apache Oozie can help implement this feature. Keep in mind that security engineers in the organization need to be tasked with examining and monitoring these files. It's important to ensure that auditing, maintaining, and analyzing logs are done consistently across the enterprise.

- Secure Configurations for Hardware and Software. Build servers based on secure images for all systems in your organization's big data architecture. Ensure patching is up to date on these machines and that administrative privileges are limited to a small number of users. Use automation frameworks, like Puppet, to automate system configuration and ensure that all big data servers in the enterprise are uniform and secure.

- Account Monitoring and Control. Manage accounts for big data users. Require strong passwords, deactivate inactive accounts, and impose a maximum permitted number of failed log-in attempts to help stop attacks from getting access to a cluster. It's important to note that the enemy isn't always outside of the organization. Monitoring account access can help reduce the probability of a successful compromise from the inside.

Organizations that are serious about big data security should consider these first steps. Cyber criminals are never going to stop being on the offensive, and with such a big target to protect, it is prudent for any enterprise utilizing big data technologies to be as proactive as possible in securing its data.

✎ New Words

challenge	[ˈtʃælindʒ]	n.挑战
		vt.向……挑战
repository	[riˈpɔzitəri]	n.知识库，仓库
aggregate	[ˈægrigeit]	v.聚集，集合，合计
		n.合计，总计，集合体
		adj.合计的，集合的，聚合的
extract	[iksˈtrækt]	vt.析取，吸取
tailor	[ˈteilə]	vt.适应，适合
		v.制作
enormous	[iˈnɔːməs]	adj.巨大的，庞大的
inefficiency	[ˌiniˈfiʃənsi]	n.无效率，无能
consolidated	[kənˈsɔlideitid]	adj.整理过的；统一的；加固的
tempting	[ˈtemptiŋ]	adj.诱惑人的
attacker	[əˈtækə]	n.攻击者
recognition	[ˌrekəgˈniʃən]	n.赞誉，承认，重视，公认，赏识
devastating	[ˈdevəsteitiŋ]	adj.破坏性的，全然的
amplify	[ˈæmplifai]	vt.放大，增强
deposit	[diˈpɔzit]	vt.存放，堆积
		vi.沉淀
		n.堆积物，存放物
dataset	[ˈdeitəset]	n.数据集
regulatory	[ˈregjulətəri]	adj.调整的
adequate	[ˈædikwit]	adj.适当的，足够的
workflow	[ˈwəːkfləu]	n.工作流
adversary	[ˈædvəsəri]	n.敌手，对手
configuration	[kənˌfigjuˈreiʃən]	n.构造，配置
authenticate	[ɔːˈθentikeit]	v.鉴别
node	[nəud]	n.节点
vulnerability	[ˌvʌlnərəˈbiləti]	n.弱点；攻击
straightforward	[streitˈfɔːwəd]	adj.坦率的，简单的，易懂的，直截了当的
		adv.坦率地
patch	[pætʃ]	n.补丁
		vt.打补丁，修补
automation	[ˌɔːtəˈmeiʃən]	n.自动控制，自动操作

framework	['freimwə:k]	n.构架，框架，结构
uniform	['ju:nifɔ:m]	adj.统一的，相同的，一致的
deactivate	[di:'æktiveit]	vt.使无效，使不活动
inactive	[in'æktiv]	adj.不活动的，停止的
probability	[,prɔbə'biliti]	n.可能性，或然性，概率
offensive	[ə'fensiv]	adj.讨厌的，无礼的，攻击性的 n.进攻，攻势
prudent	['pru:dənt]	adj.谨慎的

Phrases

consumer need	客户需求，消费者的要求
share with	分享，分与，分派
crown jewels	核心业务，顶尖业务
trade secret	商业秘密，行业秘密
upwards of	以上；多于
financial institution	金融机构
government regulation	政府管制，政府法规
come into play	开始活动
on a case-by-case basis	按照具体问题具体分析原则
data transfer	数据传送
distributed environment	分布环境
programming tool	程序设计工具
role-based access control	基于角色的访问控制
on the fly	即时
proprietary technology	专利技术
application layer	应用层
permit of	允许
cyber criminal	计算机犯罪
as proactive as possible	尽可能主动

Abbreviations

PII (Personally Identifiable Information)	个人身份信息
SANS (SysAdmin, Audit, Network, Security)	系统管理、稽核、网络及安全

Notes

[1] Big data provides an enormous competitive advantage for corporations, helping businesses tailor their products to consumer needs, identify and minimize corporate inefficiencies, and share data with user groups across the enterprise.

本句中，helping businesses tailor their products to consumer needs, identify and minimize corporate inefficiencies, and share data with user groups across the enterprise 是一个动名词短语，对 an enormous competitive advantage 进行补充说明。

[2] A breach in a big data repository could be even more damaging at a financial institution or healthcare provider, where the value of the data is extremely high and government regulations come into play.

本句中，where the value of the data is extremely high and government regulations come into play 是一个非限定性定语从句，修饰和限定 at a financial institution or healthcare provider。

[3] The variety, velocity and volume of big data amplify the security management challenges that are addressed in traditional security management.

本句中，that are addressed in traditional security management 是一个定语从句，修饰和限定 the security management challenges。

[4] With a large number of servers, there is an increased possibility that the configuration of servers may not be consistent and that certain systems may remain vulnerable.

本句中，that the configuration of servers may not be consistent and that certain systems may remain vulnerable 是 and 连接的两个并列句，作 possibility 的同位语，对其进行补充说明。

Exercises

【Ex. 1】 根据课文内容回答问题。

1. What are more and more companies doing as the amount of data being collected continues to grow?
2. What amplify the security management challenges that are addressed in traditional security management?
3. What may a big data environment include?
4. What do physical security controls need to be when big data environments are distributed geographically?
5. What is an additional big data security challenge?
6. What is the advantage of NoSQL?

7. What are the controls that the author would recommend to address the security challenges presented by big data?
8. What do proprietary technologies like Cloudera Sentry or DataStax Enterprise offer?
9. What do security engineers in the organization need to be tasked with?
10. What can monitoring account access do?

【Ex. 2】 把下列句子翻译为中文。
1. In this case, you can create a data source that retrieves data from the work items in the repository.
2. This is a quick way to have a form that you can edit and tailor according to your requirements.
3. An attacker who successfully exploited this vulnerability could run arbitrary code as the logged-on user.
4. We are using this transistor to amplify a telephone signal.
5. The dataset might also contain another table with order information.
6. This establishes a workflow between use cases.
7. Consequently, each registered base node might have different user registries configured if security is enabled.
8. Older machines will need a software patch to be loaded to correct the date.
9. Administrative staff may be deskilled through increased automation and efficiency.
10. At this point, you may activate or deactivate whatever other plugins you wish.

【Ex. 3】 短文翻译。

Big Data is the current buzzword in the technology sector, but in fields such as security it is much more than this. Businesses are starting to bet strongly on the implementation of tools based on the collection and analyzing of large volumes of data to allow them to detect malicious activity. What started out at a fashionable term has turned into a fundamental part of how we operate.

So, what exactly are the advantages of Big Data? Well, have a think about the current situation in which the use of mobile devices is growing, the Internet of Things has arrived, the number of Internet users is reaching new highs, and quickly you realize that all of this is prompting an increase in the number of accesses, transactions, users, and vulnerabilities for technology systems. This results in a surge in raw data (on the World Wide Web, on databases, or on server logs), which is increasingly more complex and varied, and generated rapidly.

Given these circumstances, we are encouraged to adopt tools that are capable of capturing and processing all of this information, helping to visualize its flow and apply automatic learning techniques that are capable of discovering patterns and detecting anomalies.

【Ex. 4】将下列词填入适当的位置（每词只用一次）。

| monitoring | exposure | relieves | reputational | mobile |
| requirements | organization | storage | independently | growing |

Key Challenges for Big Data Security

- Cyber Criminals. As it becomes bigger and more difficult to manage, big data consequently becomes more appealing to hackers and cyber criminals. Because big data is a dataset of unprecedented size with centralized access, any __(1)__ is total exposure. These types of breaches make headlines, incite consumers, and may cause major __(2)__, legal, and financial damage.
- Resource Capacity. As an organization collects big data across channels at an exponential rate, their __(3)__ can grow beyond terabytes. As a result, data encryption and migration can get bottle-necked or leaky. Additionally, the sheer volume of data makes implementation of security control unwieldy. The tools required for __(4)__ and analyzing big data produce massive amounts of their own security-related data every day, which puts undue pressure on the organization's capacity to store and analyze it all.
- Cloud and Remote Access. One answer to the capacity issues of big data is to put it in the cloud. This __(5)__ some of the burden for storage and processing, but creates new challenges for protecting it from criminals. And as more businesses allow for flex-time and __(6)__ offices, employees have access to sensitive company data via smart phones, tablet devices, and home laptops. Protecting personal devices becomes a balancing act between security and productivity.
- Supply Chain and Partner Security. Organizations rarely operate __(7)__. They rely on supply chain partners and external vendors for many of their business functions. Information flows in and out of each __(8)__ to keep these relationships functioning. Coordinating the safety of big data across partners is another layer of complexity to a business's information security challenges.
- Privacy. Both private and public organizations face the __(9)__ challenge of privacy concerns. Consumers are wary about personal information being collected and stored, and fearful about security breaches. Plus, there are legislative and regulatory __(10)__ to keep in mind.

Text B

The Future of Big Data—Big Data 2.0

For data geeks like myself, it has been a hell of a ride. The rise of big data in marketing and media has brought great interest and excitement to people. Finally, the creative directors, C-suite, and account leaders are leaning on the data scientists once again to provide deep consumer understanding and insights that are backed up and proven by actual consumers.

Today, clients often ask me about the future of big data and what the next step is; how can we leverage data on an even deeper level in order to extract meaningful consumer insights that go beyond where we are now? Most of the standard answers are around the ability to get data and insights in real time and from more devices than ever. While it is true that the connected homes, wearables, and connected cars will allow us to collect a much wider set of data points, I believe that this is just an extension of the existing approach.

It's time we move beyond structured data and into the prime time of text analytics. Here's why.

1. Numeric vs. Emotional

Most of the data points collected today are numerical or binary. They tell us if somebody engaged with a site, how well, how long, and where they engaged, but the data fails to tell us why. I believe the future of big data—Big Data 2.0 (to coin a term)—is not about more binary and numeric data points, but instead about asking the deeper questions. Big Data 2.0 should be focused not on what and where but on answering why. It should be concerned with getting a better understanding of the consumer's emotional state and the decision logic, and thereby provide deeper insight into the consumers' choices. If we focus on why instead of how often, we can create more meaningful, quality connections between consumers and brands. In other words, while numbers are great indicators of performance, focusing solely on them means brands miss the element of human connection.

Take Amazon data as an example. Amazon is filled with great numerical indicators. Its data can tell us the sales ranks (how many sold relative to category), the customer engagement (how many people shared product reviews), and their satisfaction with the product (the positive and negative reviews). All of these are great indicators, but they are still very simple and only tell a small part of the story.

Let's assume we are a consumer packaged goods company and we want to introduce a new line of diapers into the market. We decide to look at Amazon in order to better understand which products are category leaders (sales rank and number of sales) and how the consumers like the product itself (reviews). If we analyze these metrics across all diapers, we have a Big Data 1.0 picture that tells us exactly who sells the most and what the audience favorite is.

This is not enough anymore; Big Data 2.0 needs to be about the why: Why is a particular product the most sold? Why does it have an average rating of 5?

2. What's the Solution?

For us, the easiest way to get started with Big Data 2.0 is to focus on the unstructured data we collect every day. This can be reviews, customer support emails, community forums, even your own CRM system. The simplest way to look at this data is through a process called text analytics.

Text analytics is a fairly straightforward process that breaks out like this:

(1) Acquisition: Collecting and aggregating the raw data you want to analyze

(2) Transforming & Preprocessing: Cleaning and formatting the data to make it easier to read

(3) Enrichment: Enhancing the data by adding additional data points

(4) Processing: Performing specific analyses and classifications on the data

(5) Frequencies & Analysis: Evaluation of the results and translation into numerical indicators

(6) Mining: Actual extraction of information

3. Real-World Uses

Here's a real-world application using our example above. We are trying to understand the diaper market. In order to not turn this into a step-by-step guide, let's assume that we already have collected all diapers reviews as well as their qualitative indicators. That means we know what sells best and what ranks best/worst. In order to take this to the next level, we would start to extract words and phrases from the reviews. This will tell us some of the recurring patterns and their frequencies within the reviews. I actually performed this analysis by evaluating thousands of reviews and found three very actionable insights we would have never gotten to without text analytics.

3.1 Why did it sell so well?

When I looked at the reviews of the top-selling product, I found that the most mentioned terms across the majority of the helpful reviews were "price," "special," and "value." This tells us that people did not buy it because of its quality or features, but because of its pricing. So when we are launching our product, we want to look at this one for price/value guidance instead of features.

3.2 Why didn't people like it?

This one was very revealing. The brand with the most negative reviews had an extremely high frequency around the terms "tape," "stick," "stay closed," and "open." After a few reads, I discovered that consumers had no issues with the usual key features on a diaper such as "absorbency," "leakage," or "softness," but actually had issues with the tape on the side of the diaper, and the fact that it kept opening. The amount of negative reviews overall that mentioned these issues makes us believe that this is a feature that brands don't talk about but consumers care about.

3.3 Smart filtering

One interesting issue we came across is the fact that a lot of negative reviews were not actually about the product but rather focused on shipping, stock level, and packaging concerns. By tagging and removing these from the set, we are able to evaluate purely on a product level in order to focus on product-related concerns. If we were to list our diaper on Amazon, we would recommend adding a shipping and stock level guarantee prominently in the copy – a competitive advantage that speaks directly to consumer concerns.

3.4 What do they want?

From an R&D perspective, this insight is worth gold. By evaluating reviews that have terms like "I wish," "hope," or "they should," we are able to detect common features consumers are looking for when thinking about diapers. These are great insights that address the constantly changing need of the consumers. We can feed these product feature-specific insights to our R&D team as well as our copywriters.

As you can see, when analyzing the diaper category just on Amazon alone, Big Data 2.0 yielded insights beyond binary performance indicators. We could see the crowd favorites but did not (yet) know the "why" behind purchases, or understand the positive or negative reviews until our text analytics exercise. There are countless consumer insights to be mined from textual, unstructured data that give us the voice of the consumer, their motivations, and a deeper understanding of their purchasing behavior.

I hope the above examples and thoughts would give you some good ideas and inspiration on how to think about text analytics for your organization and projects. Start looking at your existing data, export your CRM, examine your comments on your website or products mentioned in topic forums – even emails from your sales department's inbox. It's Big Data

2.0 time and that's where you'll find the gold.

🐾 New Words

geek	[gi:k]	n.极客
insight	['insait]	n.洞察力，见识
leverage	['li:vəridʒ]	n.杠杆作用
		v.为……融资，利用……牟利
wearable	['wɛərəbl]	adj.可穿用的，可佩戴的
collect	[kə'lekt]	v.收集，聚集，集中，搜集
extension	[iks'tenʃən]	n.扩展，延长
		n.扩展名
emotional	[i'məuʃənəl]	adj.情绪的，情感的
binary	['bainəri]	adj.二进制的，二进位的
engaged	[in'geidʒd]	adj.使用中的
indicator	['indikeitə]	n.指标；指示器，指示符
rank	[ræŋk]	n.排行
		vt.排列，归类于，把……分等级
positive	['pɔzitiv]	adj.肯定的，积极的
negative	['negətiv]	adj.否定的，消极的
forum	['fɔ:rəm]	n.论坛
acquisition	[ˌækwi'ziʃən]	n.收集，收获
formatting	['fɔ:mætiŋ]	n.格式化
enrichment	[in'ritʃmənt]	n.浓缩
frequency	['fri:kwənsi]	n.频率，发生次数
qualitative	['kwɔlitətiv]	adj.定性的
actionable	['ækʃənəbl]	adj.可行动的，可执行的
revealing	[ri'vi:liŋ]	adj.有启迪作用的；给人启发的；透露真情的
absorbency	[əb'sɔ:bənsi]	n.吸收性，吸收能力
leakage	['li:kidʒ]	n.漏，泄漏，渗漏
softness	['sɔftnis]	n.柔和，柔软
guarantee	[ˌgærən'ti:]	n.保证，保证书，担保
		vt.保证，担保
competitive	[kəm'petitiv]	adj.竞争的
copywriter	['kɔpiraitə]	n.广告文案作者，广告写手
countless	['kautlis]	adj.无数的，数不尽的
motivation	[ˌməuti'veiʃən]	n.动机

Phrases

a hell of a	（用来加重语气）极恶劣的，不像样的，使人受不了的
C-suite	C 型雇员，指企业最高管理层。因其英文名称开头字母都带 C，因而得名。
back up	支持
real time	实时
prime time	黄金时间，黄金时段
engage with …	与……接洽，从事
coin a term	创造一个词汇
be concerned with	关心，关注
emotional state	情绪状态
decision logic	决策逻辑
be filled with	充满着
customer engagement	客户互动
product review	产品评价，商品评论
community forum	社区论坛，社团论坛
text analytic	文本分析
raw data	原始数据
actionable insight	可执行的结论
smart filtering	智能过滤
competitive advantage	竞争优势
sales department	销售部，营业部

Abbreviations

CRM (Customer Relationship Management)　　　　客户关系管理

Exercises

【Ex. 5】 根据课文内容回答问题。

1. What question do clients often ask the author?
2. What are most of the data points collected today?
3. What does the author believe?
4. What should Big Data 2.0 be focused on?
5. What can we do if we focus on why instead of how often?

6. What is the simplest way to look at this data?
7. What is text analytics?
8. What were the most mentioned terms across the majority of the helpful reviews?
9. What did the brand with the most negative reviews had an extremely high frequency around?
10. What did Big Data 2.0 do when analyzing the diaper category just on Amazon alone?

参考译文

如何管理大数据的大安全挑战

随着正在收集的数据量不断增加,越来越多的公司正在构建大数据存储库来存储和汇总数据并提取其意义。大数据为企业提供了巨大的竞争优势,帮助企业根据消费者需求定制产品,识别并最大限度地减少企业低效率行为,并实现企业中用户群体共享数据。仅 2017 年其增长率就达到了 58%,这些技术及其好处将继续显现。

不幸的是,并不是只有合法组织在壮大。整理好的大数据集对网络攻击者极具诱惑。破解组织的大数据库可以为犯罪集团提供更大的回报。当攻击者将目光瞄准大型数据存储库时,对受影响的组织可能带来毁灭性的后果。这些存储库中的数据可能包括公司的核心机密客户数据、员工数据和商业秘密。最近的目标公司数据泄露估计会导致公司损失高达 11 亿美元,而 PlayStation 的数据泄露估计给索尼公司造成了 1.7 亿美元的损失。大数据库中的数据泄露可能会给金融机构或医疗保健提供者造成更大破坏,因为其数据价值极高,政府也实施了一些法律规章。

1. 数据

大数据的种类、速度和数量使传统安全管理面临更大的挑战。大型数据存储库将存储企业内各种来源的信息。这些数据使安全访问管理面临严峻挑战。每个数据源将有自己的访问限制和安全策略,这使得平衡所有数据源的安全性尤为困难,特别是在需要从数据中聚合和提取有意义的信息更为不易。例如,大数据环境可能包括专有研究信息的数据集、需要遵守法规的数据集以及个人身份信息(PII)的单独数据集。研究人员可能希望将其研究与包括个人身份信息的数据集相关联,但是应该采取什么限制措施来确保充分的安全性?保护大数据需要根据具体情况按照安全要求进行综合平衡。

此外,许多存储库从许多不同的数据源以高容量和速度收集数据,并且它们都可能具有自己的数据传输工作流程。与多个存储库的连接可能增加对手的攻击面。从 20 个不同的数据源接收馈送的大数据系统可以向攻击者提供 20 个可行的向量来尝试访问数据集群。

2. 基础设施

另一个大数据挑战是分布式的大数据环境。与单一高端数据库服务器相比，分布式环境更加复杂，也易受攻击。当大数据环境分布在不同的地理位置时，需要在所有可访问的位置进行标准化的物理安全控制。当组织中的数据科学家希望访问信息时，边界保护变得非常重要和复杂，以便确保用户访问和系统免受可能的攻击。使用大量的服务器时，服务器的配置可能不一致——其中的某些系统可能易受攻击。

3. 技术

还有一个大数据安全挑战是大数据编程工具，包括 Hadoop 和 NoSQL 数据库，它们最初并没有考虑安全性。例如，Hadoop 原来没有对服务或用户进行身份验证，并且没有对在环境中的节点之间传输的数据进行加密。这会产生身份验证和网络安全漏洞。NoSQL 数据库缺少传统数据库提供的一些安全功能，例如基于角色的访问控制。NoSQL 的优点在于它的灵活性，允许包括新的数据类型，但是这些技术并没有为这些新数据制定安全策略。

4. 保护大数据

如何把传统数据库管理的安全性带到大数据中？几个组织描述和定义了不同的安全控制。SANS 研究所提供了 20 项安全控制。下面列出解决大数据提供的安全挑战的几个控件。

- 应用软件安全性。使用开源软件的安全版本。如上所述，大数据技术最初并没有考虑安全性。使用 Apache Accumulo 或 .20.20x 版本的 Hadoop 或更高版本等开源技术可以应对这一挑战。此外，像 Cloudera Sentry 或 DataStax Enterprise 这样的专有技术在应用层提供了增强的安全性。具体来说，Sentry 和 Accumulo 还支持基于角色的访问控制，以增强 NoSQL 数据库的安全性。
- 审核日志的维护、监控和分析。实施审计记录技术来了解和监控大型数据集群。像 Apache Oozie 这样的技术可以帮助实现这一功能。请记住，组织中的安全工程师需要负责检查和监视这些文件。确保在整个企业中一致地进行审计、维护和分析日志，这非常重要。
- 硬件和软件的安全配置。基于组织大数据架构中所有系统的安全映像构建服务器。确保在这些计算机上及时更新补丁，并且只给少量用户管理权限。使用像 Puppet 这样的自动化框架来自动化系统配置，并确保企业中的所有大型数据服务器是统一和安全的。

- 账户监控。管理大数据用户的账户。需要强大的密码，关闭不活动的账户，并利用最大允许的登录失败次数来阻止对群集攻击。尤为重要的是，要注意敌人并不总是在组织之外。监控账户访问可以降低内部威胁的可能性。

对注重大数据安全的组织来说，应该考虑这些控件。网络犯罪分子永远不会停止进攻，而且如果有这么大的保护目标，任何利用大数据技术的企业都应该尽可能地保护数据的安全。

图书资源支持

感谢您一直以来对清华版图书的支持和爱护。为了配合本书的使用,本书提供配套的资源,有需求的读者请扫描下方的"书圈"微信公众号二维码,在图书专区下载,也可以拨打电话或发送电子邮件咨询。

如果您在使用本书的过程中遇到了什么问题,或者有相关图书出版计划,也请您发邮件告诉我们,以便我们更好地为您服务。

我们的联系方式:

地　　址: 北京市海淀区双清路学研大厦 A 座 701

邮　　编: 100084

电　　话: 010-62770175-4608

资源下载: http://www.tup.com.cn

客服邮箱: tupjsj@vip.163.com

QQ: 2301891038(请写明您的单位和姓名)

用微信扫一扫右边的二维码,即可关注清华大学出版社公众号"书圈"。

资源下载、样书申请

书圈

扫一扫,获取最新目录